Lecture Notes Editorial Policies

Lecture Notes in Statistics provides a format for the informal and quick publication of monographs, case studies, and workshops of theoretical or applied importance. Thus, in some instances, proofs may be merely outlined and results presented which will later be published in a different form.

Publication of the Lecture Notes is intended as a service to the international statistical community, in that a commercial publisher, Springer-Verlag, can provide efficient distribution of documents that would otherwise have a restricted readership. Once published and copyrighted, they can be documented and discussed in the scientific literature.

Lecture Notes are reprinted photographically from the copy delivered in camera-ready form by the author or editor. Springer-Verlag provides technical instructions for the preparation of manuscripts. Volumes should be no less than 100 pages and preferably no more than 400 pages. A subject index is expected for authored but not edited volumes. Proposals for volumes should be sent to one of the series editors or addressed to "Statistics Editor" at Springer-Verlag in New York.

Authors of monographs receive 50 free copies of their book. Editors receive 50 free copies and are responsible for distributing them to contributors. Authors, editors, and contributors may purchase additional copies at the publisher's discount. No reprints of individual contributions will be supplied and no royalties are paid on Lecture Notes volumes. Springer-Verlag secures the copyright for each volume.

Series Editors:

Professor P. Bickel
Department of Statistics
University of California
Berkeley, California 94720
USA

Professor P. Diggle
Department of Mathematics
Lancaster University
Lancaster LA1 4YL
England

Professor S. Fienberg
Department of Statistics
Carnegie Mellon University
Pittsburgh, Pennsylvania 1521
USA

Professor K. Krickeberg
3 Rue de L'Estrapade
75005 Paris
France

Professor I. Olkin
Department of Statistics
Stanford University
Stanford, California 94305
USA

Professor N. Wermuth
Department of Psychology
Johannes Gutenberg Universit
Postfach 3980
D-6500 Mainz
Germany

Professor S. Zeger
Department of Biostatistics
The Johns Hopkins University
615 N. Wolfe Street
Baltimore, Maryland 21205-2
USA

Lecture Notes in Statistics 180

Edited by P. Bickel, P. Diggle, S. Fienberg, U. Gather,
I. Olkin, S. Zeger

Lecture Notes in Statistics

For information about Volumes 1 to 126, please contact Springer-Verlag

Yanhong Wu

Inference for Change-Point and Post-Change Means After a CUSUM Test

 Springer

Yanhong Wu
Department of Mathematics
University of the Pacific
Stockton, CA 95211
ywu@pacific.edu

Library of Congress Cataloging-in-Publication Data
Wu, Yanhong.
    Inference for change-point and post-change means after a CUSUM test / Yanhong Wu.
      p. cm. — (Lecture notes in statistics, ISSN 0930-0325 ; 180)
    Includes bibliographical references and index.
    1. Statistics.   2. Probabilities.   3. Stochastic processes.   4. Econometrics.   I. Title.   II.
Lecture notes in statistics (Springer-Verlag) ; v. 180.
    QA276.W82   2004
    519.5—dc22   2004058917

ISBN 0-387-22927-2   Printed on acid-free paper.

Camera ready copy provided by the editors.

Printed in the United States of America.   (SB)

9 8 7 6 5 4 3 2 1   SPIN 11015413

springeronline.com

Preface

The change-point problem has attracted many statistical researchers and practitioners during the last few decades. Here, we only concentrate on the sequential change-point problem. Starting from the Shewhart chart with applications to quality control [see Shewhart (1931)], several monitoring procedures have been developed for a quick detection of change. The three most studied monitoring procedures are the CUSUM procedure [Page (1954)], the EWMA procedure [Roberts (1959)] and the Shiryayev–Roberts procedure [Shiryayev (1963) and Roberts (1966)]. Extensive studies have been conducted on the performances of these monitoring procedures and comparisons in terms of the delay detection time. Lai (1995) made a review on the state of the art on these charts and proposed several possible generalizations in order to detect a change in the case of the unknown post-change parameter case. In particular, a window-limited version of the generalized likelihood ratio testing procedure studied by Siegmund and Venkatraman (1993) is proposed for a more practical treatment even when the observations are correlated.

In this work, our main emphasis is on the inference problem for the change-point and the post-change parameters after a signal of change is made. More specifically, due to its convenient form and statistical properties, most discussions are concentrated on the CUSUM procedure. Our goal is to provide some quantitative evaluations on the statistical properties of estimators for the change-point and the post-change parameters. It has to be stressed that there have been many studies on the inference problem for the change-point in the fixed sample size case under both parametric models and nonparametric models. The inference problem after sequential detection raises both theoretical and technical difficulties, as mentioned in a preliminary study by Hinkley (1971). From the theoretical point of view, due to the irregularity of the change-point problem, it will be interesting to see how the sequential sampling plan affects the statistical properties. Interestingly, we shall see that even in the normal case, the bias of the change-point estimator is different from the fixed sample size case even in the large sample situation. From the technical point of view, due to the nature of sequential sampling plan, the inference problem for the change-point and the post-change parameters raises more technical difficulties compared with other standard situations. The inference problem for the post-change parameters involves more careful treatments because of the uncertainty of the change-point estimator. From the application point of view, the change-point problem has been raised from a variety of areas where changes in structures of dynamic systems are the interest such as online quality control, financial market and economic systems, global warming and DNA sequence analysis, to mention a few. Motivated by these reasons, the author feels that it may attract researchers' and practitioners' interest from the results and techniques developed in this work as well as from several case studies.

The notes are organized as follows. In Chapter 1, we first introduce the regular CUSUM procedure and give a simple approximation for the average run lengths for design purposes. Then we explain the strong renewal theorem in the exponential family case. As a demonstration, we give a derivation for the approximation for the in-control average run length in the normal case. The estimators for the change-point and the post-change parameter are proposed.

In Chapter 2, we consider the bias and the absolute bias for the estimator of the change-point conditioning on a change that is detected. By assuming both the change-point and the threshold approach infinity, the asymptotic quasi-stationary bias is derived. In order to study the local properties, we further assume both the reference value and the post-change mean approach zero at the same order and obtain the second-order approximation for the bias. It is shown that in contrast to the fixed sample size case as considered in Wu (1999), the asymptotic bias is not zero even in the normal distribution case.

In Chapter 3, we propose a method of constructing a lower confidence limit for the change-point. This may be of interest from a quality inspection point of view, where we need to estimate the number of items to be inspected after a change is detected.

In Chapter 4, we concentrate on the inference problem for the post-change mean in the normal case in order to develop the technique. The bias and a corrected normal pivot based on the post-change mean estimator are studied. The techniques developed here provide new methods of studying the change-point problem.

In Chapter 5, we study the behavior of the post-change mean estimator when the signal is false. This serves two purposes. First, from theoretical aspects, the post-change mean estimator under a false signal can be treated as the sample mean for a conditional random walk with negative drift by staying positive. The results extend the available asymptotic theory in the sense that we obtain the second-order correction for the normal approximation. Second, by studying the behavior of the post-change mean under a false signal, it provides a method of separating the regular clusters from sparse changing segments.

In Chapter 6, we extend the results to a specific problem in the normal case when the variance (nuisance parameter) is subject to possible changes as well. In other words, we want to study the inference problem for the change-point and the post-change mean when the post-change variance needs to be estimated. This provides a robust analysis for the CUSUM procedure for monitoring changes in mean. We find that more technical difficulties are raised, particularly for constructing the confidence interval for the post-change mean.

In Chapter 7, we consider a sequential classification and segmentation procedure for alternatively changing means as an application of the results in Chapter 2. By using the dam process, we can classify alternatively changing means sequentially and also locate the segments after correction in the offline case. The method provides an alternative to the Bayesian sequential filtering procedure.

In Chapter 8, in order to deal with more complex post-change parameters such as linear post-change means in the normal case, we propose an adaptive CUSUM procedure. The adaptive CUSUM procedure uses a sequence of adaptive sequential tests with a two-sided stopping boundary. The advantage for this adaptive CUSUM procedure is its convenience of locating the change-point and estimating the post-change parameters. Theoretical properties of the procedure are studied under the simple change model. The application to global warming detection is used for an illustration.

In Chapter 9, we generalize the methods to a correlated data case. In particular, we consider the model-based CUSUM procedure when the observations follow an AR(1) model with mean subject to change. We study the effect of correlation on the biases of change-point estimator and post-change mean estimator. Simulation study for the effect of correlation on the classical CUSUM procedure is also carried out.

In Chapter 10, we consider the estimator for the change-point under the Shiryayev−Roberts procedure. A comparison with the CUSUM procedure is conducted. Some concluding remarks on the generalizations to other models including the multi-variate observation case and multidimensional parameter case are made with several possible open problems posted.

Readers with basic statistical training in sequential analysis with an elementary probability background should be able to understand most results. Wald's likelihood ratio identity [Wald (1947)] is used extensively. A convenient reference is Siegmund (1985). For those readers who need to review the basic properties about the CUSUM procedure, we refer to van Dobben De Bruyn (1968) and Hawkins and Olwell (1998). Most results are given under the simple change-point model; that means, the observations are independent. In fact, most discussions are concentrated on the normal distribution case, which, we hope, will attract most researchers and practitioners who are interested in change-point problems. The techniques and methods are nevertheless quite general and can be extended directly to the more general exponential family case and dependent observation models such as Markovian structures. Also to make the work more attractive to practitioners, several case studies are added at the end of each major chapter; these case studies show the detailed procedures of applying the methods and results to the real data sets. Original data sets are listed to form a relatively self-contained package.

The majority of the results in the notes were derived when the author was visiting the Department of Statistics at University of Michigan which has an encouraging and stimulating environment. Frequent discussions with the faculty there, particularly Professors M. Woodroofe and R. Keener, were very helpful. The writing was finished when the author was currently visiting the Department of Mathematics at the University of the Pacific, and their hospitality is greatly appreciated. Editorial assistance from Dr. J. Kimmel, editor of the series, A. Orrantia from the production department, and many comments from the series editors and two reviewers helped me to provide a much improved revision. The long-time support from my family makes the writing and research possible. This book is dedicated to my late father.

Yanhong Wu
September, 2004

Contents

List of Tables

1
CUSUM Procedure

1.1   CUSUM Procedure for Exponential Family

Let $F_\theta(x)$ belong to a standard one-parameter exponential family of the form

$$dF_\theta(x) = \exp(x\theta - c(\theta))dF_0(x),$$

for $|\theta| \leq K(>0)$ and $c(\theta)$ is the cumulant generating function such that

$$c(0) = c'(0) = 0, \quad c''(0) = 1.$$

Also denote by

$$\gamma = c^{(3)}(0) \quad \text{and} \quad \kappa = c^{(4)}(0)$$

the third and fourth cumulants of $F_0(x)$. Throughout our discussion, we shall assume that $F_0(x)$ is strongly nonarithmetic in the sense that

$$\lim_{|\lambda|\to\infty} \sup \left| \int_{-\infty}^{\infty} e^{i\lambda x} dF_0(x) \right| < 1.$$

This condition implies that $F_\theta(x)$ is also strongly nonarithmetic uniformly for $|\theta| < \theta^*$ for some $\theta^* > 0$ [Siegmund (1979)]. When $F_0(x)$ has the density $f_0(x)$, we denote by $f_\theta(x)$ the density of $F_\theta(x)$.

Suppose $\{X_k\}$ are independent random variables that follow distribution $F_{\theta_0}(x)$ for $k \leq \nu$ and $F_\theta(x)$ for $k > \nu$, where $\theta_0 < 0 < \theta$ and ν is the change-point. For a preselected reference value $\theta_1 > 0$ for θ, which is the conjugate pair for θ_0 such that

$$c(\theta_0) = c(\theta_1),$$

the log-likelihood ratio function at $\nu = k$ for $0 \leq k \leq n$ versus $\nu = n$ for given $X_1, ..., X_n$ is

$$l(k) = (\theta_1 - \theta_0)(X_{k+1} + ... + X_n),$$

where $l(n) = 0$. Thus, the generalized likelihood ratio test in the fixed sample size case is based on

$$
\begin{aligned}
\max_{0 \leq k \leq n} l(k) \propto T_n &= \max_{0 \leq k \leq n} (X_{k+1} + ... + X_n) \\
&= \max(0, T_{n-1} + X_n),
\end{aligned}
$$

with $T_0 = 0$.

The CUSUM procedure [Page (1954)], based on the sequential likelihood ratio test, signals a change at the time

$$N = \inf\{n > 0: \ T_n > d\},$$

where d is the threshold. At $\theta = \theta_1$, the maximum likelihood estimator of ν conditional on the change being detected is given by

$$\hat{\nu} = \max\{k < N: \ T_k = 0\},$$

which is the last zero point of T_k before the time to signal.

To estimate the post-change parameter, we hope that the change-point estimator $\hat{\nu}$ is not too far from the true change-point ν. Thus, the post-change parameter θ is estimated from the following equation:

$$c'(\hat{\theta}) = \frac{1}{N - \hat{\nu}} \sum_{i=\hat{\nu}+1}^{N} X_i,$$

where $c'(\theta)$ is the mean of X_i's under $P_\theta(.)$.

This work focuses on the inference problem for ν and θ based on $\hat{\nu}$ and $\hat{\theta}$. In this chapter, we first present some preliminary results and a basic background on the CUSUM procedure.

1.2   Operating Characteristics

In this section, we consider the evaluation of operating characteristics of the CUSUM procedure and consequently study the design problems for the related parameters.

For the convenience of discussion throughout the whole notes, we first introduce some notations. Define

$$S_n = S_0 + \sum_{i=1}^{n} X_i.$$

For general $S_0 = x$ with $0 \leq x < d$, we define

$$N_x = \inf\{n > 0 : \ S_n \leq 0 \ \text{ or } \ > d\},$$

as the two-sided boundary crossing time for S_n with boundary $[0, d)$. In the particular case of $S_0 = 0$, we define

$$\tau_y = \inf\{n > 0 : \ S_n > (\leq)y\}, \ \text{ for } \ y > (\leq)0,$$

as the boundary crossing time with boundary y and

$$R_y = S_{\tau_y} - y$$

as the overshoot and

$$R_{\pm\infty} = \lim_{y \to \pm\infty} R_y$$

as the limit of overshoot when it exists. In particular, let

$$\tau_- = \tau_0 \ \text{ and } \ \tau_+ = \lim_{y \to 0+} \tau_y$$

denote the ladder epochs, which denote the first time the random walk goes below zero or above zero.

The CUSUM procedure can be interpreted as a sequence of sequential tests with two-sided boundary $[0, d)$ [for example, Siegmund (1985, page 24)]. Define $P_{\theta_0}(.)$ and $P_\theta(.)$ as the measures with parameters θ_0 and θ, respectively, and $P^\nu(.)$ the measure when the change is at ν. Throughout our discussion, we shall denote by $E[X; A] = E[X I_A]$ the expectation when the event A occurs where I_A is the indicator function for A and $E[X|A]$ as the conditional expectation of X given A. As a convention, we see that

$$E[X; A] = E[X|A]P(A).$$

Note that

$$N_0 = \inf\{n > 0 : \ S_n \leq 0 \ \text{ or } \ > d\},$$

when $x = 0$ for N_x. Thus, N can be described as a random sum of stopping times, which are equivalent to N_0 in distribution, until the first time a sequential test crosses d. Using Wald's identity, we have

$$ARL_0 = E_{\theta_0}(N) = \frac{E_{\theta_0}(N_0)}{P_{\theta_0}(S_{N_0} > d)}$$

and

$$ARL_1 = E_{\theta_1}(N) = \frac{E_{\theta_1}(N_1)}{P_{\theta_1}(S_{N_0} > d)}.$$

The design for the monitoring limit d is typically based on ARL_0. The evaluation of ARL_0 and ARL_1 can be done by either a simulation or numerical method; see van Dobben de Bruyn (1968), Brook and Evans (1973), Reynolds (1975), Woodall (1983) and more recently Hawkins and Olwell (1998). Here we present the following simple local approximation [Siegmund (1985, Chap.2)] by first assuming $d \to \infty$ and then letting θ_0, θ_1 approach zero:

$$ARL_0 \approx \frac{e^{(\theta_1-\theta_0)(d+\rho_+-\rho_-)} - 1 - (\theta_1 - \theta_0)(d + \rho_+ - \rho_-)}{-(\theta_1 - \theta_0)c'(\theta_0)}$$

and

$$ARL_1 \approx \frac{e^{-(\theta_1-\theta_0)(d+\rho_+-\rho_-)} - 1 + (\theta_1 - \theta_0)(d + \rho_+ - \rho_-)}{(\theta_1 - \theta_0)c'(\theta_1)},$$

where

$$\rho_+ = E_0 R_\infty \quad \text{and} \quad \rho_- = E_0 R_{-\infty}.$$

In Section 1.4, we show that the above approximations are, indeed, accurate up to the second-order.

More realistic measures for the average delay detection time can be defined as

$$\lim_{\nu \to \infty} E^\nu[N - \nu | N > \nu]$$

or

$$\text{esssup}_\nu E^\nu[N - \nu | N > \nu].$$

The former is defined if the change occurs far away from the starting time, while the latter takes the worst situation no matter where the change occurs. Lorden (1971), Moustakides (1986) and Ritov (1991) showed the optimality of the CUSUM procedure under the latter measure. Detailed discussion and comparisons can be seen in Shiryayev (1963), Pollak and Siegmund (1985) and Srivastava and Wu (1993). An evaluation under the former measure in the normal case is given in the appendix to Chapter 7.

1.3 Strong Renewal Theorem and Ladder Variables

In this section, we state the strong renewal theorem and its applications to the approximations for the probability and moments associated with the ladder variables. The strong renewal theorem is given in Stone (1965) and developed in Siegmund (1979) in the exponential family case.

We first define the ladder variables. Let

$$\tau_-^{(0)} = \tau_+^{(0)} = 0$$

and

$$\tau_+^{(1)} = \tau_+, \quad \tau_-^{(1)} = \tau_-.$$

For $n > 1$, we define

$$\tau_+^{(n)} = \inf\{k : S_k > S_{\tau_+^{(n-1)}}\},$$

$$\tau_-^{(n)} = \inf\{k : S_k \leq S_{\tau_-^{(n-1)}}\},$$

as the nth ascent and descent ladder epoches, respectively. Denote for $x \geq 0$,

$$U_\theta^+(x) = \sum_{n=1}^\infty P_\theta(S_{\tau_+^{(n)}} \leq x), \quad U_{\theta_0}^-(x) = \sum_{n=1}^\infty P_{\theta_0}(-S_{\tau_-^{(n)}} \leq x),$$

as the corresponding renewal functions of $\{S_n\}$.

For notational convenience, we denote by

$$\Delta_0 = \theta_1 - \theta_0 \quad \text{and} \quad \Delta = \theta - \tilde{\theta},$$

where $\theta_1 > 0$ and $\tilde{\theta} < 0$ are the conjugate pairs of θ_0 and θ such that

$$c(\theta_0) = c(\theta_1) \quad \text{and} \quad c(\theta) = c(\tilde{\theta}).$$

Also, let

$$\mu = c'(\theta), \quad \tilde{\mu} = c'(\tilde{\theta}), \quad \text{and} \quad \mu_i = c'(\theta_i),$$

as the corresponding means for $i = 0, 1$.

The following uniform strong renewal theorem is stated in Chang (1992) which formalizes the results of Siegmund (1979). The proof is omitted here.

UNIFORM STRONG RENEWAL THEOREM: *There exist positive numbers r, θ^*, and C such that*

$$\left| U_\theta^+(x) - \frac{x}{E_\theta S_{\tau_+}} - \frac{E_\theta S_{\tau_+}^2}{2(E_\theta S_{\tau_+})^2} \right| \leq Ce^{-rx},$$

$$\left| U_{\theta_0}^-(x) + \frac{x}{E_{\theta_0} S_{\tau_-}} - \frac{E_{\theta_0} S_{\tau_-}^2}{2(E_{\theta_0} S_{\tau_-})^2} \right| \leq Ce^{-rx},$$

uniformly for $x \geq 0$ and

$$-\theta^* \leq \theta_0 < 0 \leq \theta \leq \theta^*.$$

The strong renewal theorem states that the error terms are at the exponential order. An immediate consequence of the strong renewal theorem is the exponential convergence rate for the overshoot, which is stated in Chang (1992).

Corollary 1.1: *There exist positive numbers r, θ^*, and C such that*

$$|P_\theta(R_x \leq y) - P_\theta(R_\infty \leq y)| \leq Ce^{-r(x+y)},$$
$$|P_{\theta_0}(R_{-x} > -y) - P_{\theta_0}(R_{-\infty} > -y)| \leq Ce^{-r(x+y)},$$

uniformly for $x, y \geq 0$ and

$$-\theta^* \leq \theta_0 \leq 0 \leq \theta \leq \theta^*,$$

where

$$P_\theta(R_\infty \leq y) = \frac{1}{E_\theta S_{\tau_+}} \int_0^y P_\theta(S_{\tau_+} > x) dx,$$

and

$$P_{\theta_0}(R_{-\infty} > -y) = \frac{1}{-E_{\theta_0} S_{\tau_-}} \int_0^y P_{\theta_0}(S_{\tau_-} < -y) dy.$$

We denote

$$\rho_{\pm}^{(k)} = E_0 R_{\pm\infty}^k \quad \text{and} \quad \rho_{\pm} = \rho_{\pm}^{(1)}$$

as the moments of limit overshoot.

One important application of the strong renewal theorem is to deliver very accurate local approximations for the (joint) moments of ladder variables when the mean approaches zero. The following lemma, which gives two very important approximations, is given in Lemma 10.27 of Siegmund (1985) and extended in Chang (1992).

Lemma 1.1: *As $0 < \theta \to 0$, for $k \geq 1$,*

$$E_\theta S_{\tau_+}^k = E_0 S_{\tau_+}^k + \frac{k}{k+1} E_0 S_{\tau_+}^{k+1} \theta + \frac{\theta^2}{2} \left(\frac{k}{k+2} E_0 S_{\tau_+}^{k+2} - \alpha_k \right) + o(\theta^2),$$

$$\mu E_\theta(\tau_+ S_{\tau_+}^k) = \frac{1}{k+1} E_0 S_{\tau_+}^{k+1} + \theta \left(\frac{1}{k+2} E_0 S_{\tau_+}^{k+2} + \alpha_k \right) + o(\theta),$$

where

$$\alpha_k = \int_{0-}^{\infty} \left(E_0 R_x^k - \rho_+^{(k)} \right) U_0^-(dx).$$

Similarly, as $0 > \theta_0 \to 0$,

$$E_{\theta_0} S_{\tau_-}^k = E_0 S_{\tau_-}^k + \frac{k}{k+1} \theta_0 E_0 S_{\tau_-}^{k+1} + \frac{\theta_0^2}{2} \left(\frac{k}{k+2} E_0 S_{\tau_-}^{k+2} - \beta_k \right)$$
$$+ o(\theta_0^2),$$

$$\mu_0 E_{\theta_0}(\tau_- S_{\tau_-}^k) = \frac{1}{k+1} E_0 S_{\tau_-}^{k+1} + \theta_0 \left(\frac{1}{k+2} E_0 S_{\tau_-}^{k+2} + \beta_k \right) + o(\theta_0),$$

where

$$\beta_k = \int_{0+}^{\infty} (E_0 R_{-x}^k - \rho_-^{(k)}) U_0^+(dx).$$

An interesting application of Lemma 1.1 is the following two different versions of approximation for the probability $P_\theta(\tau_- = \infty)$ that the random walk will never go below zero when the drift is positive.

Lemma 1.2: *As $\theta \to 0$,*

$$
\begin{aligned}
P_\theta(\tau_- = \infty) &= \frac{\mu}{E_0 S_{\tau_+}} \exp\left(-\rho_+ \theta - \frac{1}{2}\left(\rho_+^{(2)} - \rho_+^2 - \frac{\alpha_1}{E_0 S_{\tau_+}}\right)\theta^2\right)(1 + o(\theta^2)) \\
&= -\Delta E_0 S_{\tau_-} \exp\left(\rho_- \theta + \frac{1}{2}\left(\rho_-^{(2)} - \rho_-^2 - \frac{\beta_1}{E_0 S_{\tau_-}}\right)\theta^2\right) \\
&\quad \times (1 + o(\theta^2)).
\end{aligned}
$$

Proof: For the first approximation, by using the Wiener-Hopf factorization [e.g. Siegmund (1985, Corollary 8.39)] and Lemma 1.1, we have

$$
\begin{aligned}
P_\theta(\tau_- = \infty) &= \frac{1}{E_\theta(\tau_+)} \\
&= \frac{\mu}{E_\theta S_{\tau_+}} \\
&= \mu\left(E_0 S_{\tau_+} + \frac{\theta}{2} E_0 S_{\tau_+^2} + \frac{\theta^2}{2}\left(\frac{1}{3} E_0 S_{\tau_+^3} - \alpha_1\right) + o(\theta^2)\right)^{-1} \\
&= \frac{\mu}{E_0 S_{\tau_+}}\left(1 + \theta\rho_+ + \frac{\theta^2}{2}\left(\rho_+^{(2)} - \frac{\alpha_1}{E_0 S_{\tau_+}}\right) + o(\theta^2)\right)^{-1},
\end{aligned}
$$

which is equivalent to the expected result.

By using the Wald's likelihood ratio identity [Siegmund (1985, Lemma 2.24)] first by changing parameter θ to $\tilde\theta$ and then taking a local Taylor expansion, we have the following second-order approximation:

$$
\begin{aligned}
P_\theta(\tau_- = \infty) &= 1 - P_\theta(\tau_- < \infty) \\
&= 1 - E_\theta I_{[\tau_- < \infty]} \\
&= 1 - E_{\tilde\theta} e^{\Delta S_{\tau_-}} \\
&= -\left(\Delta E_{\tilde\theta} S_{\tau_-} + \frac{\Delta^2}{2} E_{\tilde\theta} S_{\tau_-}^2 + \frac{\Delta^3}{6} E_{\tilde\theta} S_{\tau_-}^3\right) + o(\Delta^3) \\
&= -\Delta\left(E_0 S_{\tau_-} + \frac{\tilde\theta}{2} E_0 S_{\tau_-}^2 + \frac{\tilde\theta^2}{2}\left(\frac{1}{3} E_0 S_{\tau_-}^3 - \beta_1\right)\right) \\
&\quad + \frac{\Delta}{2}\left(E_0 S_{\tau_-}^2 + \frac{2}{3}\tilde\theta E_0 S_{\tau_-}^3\right) + \frac{\Delta^2}{6} E_0 S_{\tau_-}^3 + o(\Delta^2)) \\
&= -\Delta E_0 S_{\tau_-}\left(1 + \theta\rho_- + \frac{\theta^2}{2}\left(\rho_-^{(2)} - \frac{\beta_1}{E_0 S_{\tau_-}}\right) + o(\theta^2)\right),
\end{aligned}
$$

which is equivalent to the expected result.

A further consequence of this lemma is the following link between the moments of X_1 and overshoot under $P_0(.)$.

By matching the two versions of approximation in Lemma 1.2, we have

$$-\frac{\Delta}{\mu} E_0 S_{\tau_+} E_0 S_{\tau_-} =$$

$$\exp\left(-\theta(\rho_+ + \rho_-) - \frac{\theta^2}{2}\left(\rho_+^{(2)} + \rho_-^{(2)} - \rho_+^2 - \rho_-^2 - \frac{\alpha_1}{E_0 S_{\tau_+}} - \frac{\beta_1}{E_0 S_{\tau_-}}\right)\right)$$
$$\times \; (1 + o(\theta^2)).$$

The Taylor expansion around zero gives

$$\mu \;=\; c'(\theta)$$
$$=\; \theta + \frac{\theta^2}{2}\gamma + \frac{\theta^3}{6}\kappa$$
$$=\; \theta \exp\left(\frac{\gamma}{2}\theta + \theta^2\left(\frac{\kappa}{6} - \frac{\gamma^2}{8}\right)\right)(1 + o(\theta^2)),$$

and

$$\Delta \;=\; 2\theta + \frac{\gamma}{3}\theta^2 + \frac{\gamma^2}{9}\theta^3$$
$$=\; 2\theta \exp\left(\frac{\gamma}{6}\theta + \frac{\gamma^2}{24}\theta^2\right)(1 + (\theta^2)).$$

Thus, we have the following useful identities:

Corollary 1.2:

$$E_0 S_{\tau_+} E_0 S_{\tau_-} \;=\; -\frac{1}{2},$$
$$\rho_+ + \rho_- \;=\; \frac{\gamma}{3};$$
$$\rho_+^{(2)} + \rho_-^{(2)} - \rho_+^2 - \rho_-^2 - \frac{\alpha_1}{E_0 S_{\tau_+}} - \frac{\beta_1}{E_0 S_{\tau_-}} \;=\; \frac{1}{3}\left(\kappa - \gamma^2\right).$$

Example 1.1: (Normal case) Here

$$c(\theta) = \frac{\theta^2}{2} \quad \text{and} \quad F_0(x) = \Phi(x),$$

the standard normal distribution, which is symmetric. Thus, $\kappa = \gamma = 0$, and from Corollary 1.2,

$$\rho_+ = -\rho_- \approx 0.5826, \quad E_0 S_{\tau_+} = -E_0 S_{\tau_-} = \frac{1}{\sqrt{2}},$$

$$\rho_+^{(2)} - \rho_+^2 - \frac{\alpha_1}{E_0 S_{\tau_+}} = \rho_-^{(2)} - \rho_-^2 - \frac{\beta_1}{E_0 S_{\tau_-}} = 0,$$

and

$$\frac{\beta_1}{E_0 S_{\tau_-}} = \rho_+^{(2)} - \rho_+^2 = \frac{1}{4}.$$

Example 1.2: (Exponential distribution)

Let $f_0(x) = e^{-(x+1)}$ for $x \geq -1$, which is standardized with mean 0 and variance 1. Then

$$c(\theta) = -\theta - \ln(1 - \theta),$$

for $|\theta| < 1$. Thus,

$$c'(\theta) = \theta/(1 - \theta) \quad \text{and} \quad c^{(k)}(\theta) = (k - 1)!/(1 - \theta)$$

for $k \geq 2$ and $\gamma = 2$ and $\kappa = 6$.

Because of the memoryless property, R_x follows $\exp(1)$ for any $x \geq 0$. Also, it is noted that S_{τ_-} follows $U(-1, 0)$ [Siegmund (1985, p.186, Problem 8.10)]. Thus,

$$E_0 S_{\tau_+} = 1, \quad \rho_+ = 1, \quad \rho_+^{(2)} = 2,$$
$$E_0 S_{\tau_-} = -\frac{1}{2}, \quad \rho_- = -\frac{1}{3}, \quad \rho_-^{(2)} = \frac{1}{6}.$$

Since $\alpha_1 = 0$, we have

$$\frac{\beta_1}{E_0 S_{\tau_-}} = \frac{7}{18}$$

from Corollary 1.2.

Example 1.3: (Chi-square distribution)

Let $f_0(x)$ be density function of $(\chi_p^2 - p)/\sqrt{2p}$, where χ_p^2 is the standard chi-square random variable with p degrees of freedom.

Then

$$c(\theta) = -\sqrt{\frac{p}{2}}\theta - \frac{p}{2}\ln\left(1 - \theta/\sqrt{\frac{p}{2}}\right).$$

Thus,

$$c'(\theta) = -\sqrt{\frac{p}{2}} + \sqrt{\frac{p}{2}} / \left(1 - \theta/\sqrt{\frac{p}{2}}\right),$$

and for $k > 1$,

$$c^{(k)}(\theta) = (k - 1)! \frac{(p/2)^{1-k/2}}{\left(1 - \theta/\sqrt{\frac{p}{2}}\right)^k}.$$

The evaluation for the quantities related to the overshoot is complicated. For example, Eq. (10.55) of Siegmund (1985) gives a formula to calculate the first moment $\rho_+ = E_0 R_\infty$ as

$$\rho_+ = \frac{\kappa}{6} - \int_0^\infty \frac{1}{\pi\lambda^2} Re\{\ln[2(1 - g(\lambda))/\lambda^2]\} d\lambda,$$

where

$$g(\lambda) = e^{c(i\lambda)}.$$

In the special case of $p = 1$, Wu and Xu (1999) shows that $\rho_+ \approx 1.20$ by a numerical integration. For general values of p, we refer to Srivastava (1997).

1.4   ARL in the Normal Case

In this section, we give a derivation of the approximation for the average run lengths in the normal case as an application of the strong renewal theorem. In the normal case, $\theta_1 = -\theta_0$. The derivation is separated into two steps. In the first step, we first derive the asymptotic result when $d \to \infty$, and in the second step, we further assume θ_0 approach zero in order to give local second-order approximation.

The first lemma gives the approximation for the terminating probability $P_{\theta_0}(S_N > d)$:

Lemma 1.4: *Uniformly for $0 \leq \theta_1 \leq \theta^*$, there exists a positive constant $r > 0$ such that*

$$P_{\theta_0}(S_N > d) = \frac{\left(E_{\theta_1} e^{2\theta_0 R_\infty}\right)^{-1} P_{\theta_1}(\tau_- = \infty)}{e^{-2\theta_0 d} \left(E_{\theta_1} e^{2\theta_0 R_\infty}\right)^{-2} - 1}(1 + o(e^{-rd})),$$

and

$$P_{\theta_1}(S_N > d) = \frac{P_{\theta_1}(\tau_- = \infty)}{1 - e^{2\theta_0 d} \left(E_{\theta_1} e^{2\theta_0 R_\infty}\right)^2} \left(1 + o(e^{-rd})\right).$$

Proof: The technique used here is very similar to the discussion in Section 8.5 of Siegmund (1985). First, we use the total probability law and Wald's likelihood ratio identity by changing parameter θ_0 to θ_1. Then we condition on the value of S_N when $S_N < 0$ and use the strong renewal theorem,

$$
\begin{aligned}
P_{\theta_0}(S_N > d) &= P_{\theta_0}(\tau_d < \infty) - P_{\theta_0}(\tau_d < \infty; S_N \leq 0) \\
&= E_{\theta_1} e^{2\theta_0 S_{\tau_d}} - E_{\theta_0}[P_{\theta_0}(\tau_d < \infty | S_N); S_N \leq 0] \\
&= (e^{2\theta_2 d} E_{\theta_1} e^{2\theta_0 R_\infty} \\
&\quad - e^{2\theta_2 d} E_{\theta_1} e^{2\theta_0 R_\infty} E_{\theta_0}\left[e^{-2\theta_0 S_N}; S_N \leq 0\right])(1 + o(e^{-rd})) \\
&= e^{2\theta_2 d} E_{\theta_1} e^{2\theta_0 R_\infty} P_{\theta_1}(S_N > d)(1 + o(e^{-rd})).
\end{aligned}
$$

On the other hand,

$$
\begin{aligned}
P_{\theta_1}(S_N < 0) &= P_{\theta_1}(\tau_- < \infty) - P_{\theta_1}(\tau_- < \infty; S_N > d) \\
&= P_{\theta_1}(\tau_- < \infty) - E_{\theta_1}[P_{\theta_1}(\tau_- < \infty | S_N); S_N > d] \\
&= P_{\theta_1}(\tau_- < \infty) - E_{\theta_1} e^{2\theta_0 R_\infty} E_{\theta_1}\left[e^{-2\theta_0 S_N}; S_N > d\right] \\
&\quad \times (1 + o(e^{-rd})) \\
&= P_{\theta_1}(\tau_- < \infty) - E_{\theta_1} e^{2\theta_0 R_\infty} P_{\theta_0}(S_N > d)(1 + o(e^{-rd})).
\end{aligned}
$$

That means,

$$P_{\theta_1}(S_N > d) = P_{\theta_1}(\tau_- = \infty) + E_{\theta_1} e^{2\theta_0 R_\infty} P_{\theta_0}(S_N > d)\left(1 + o(e^{-rd})\right).$$

Combining the above two approximations, we get the expected result.

Lemma 1.5:
$$\frac{P_{\theta_1}(\tau_- = \infty)}{E_{\theta_0} S_{\tau_-}} = -2\theta_0 E_{\theta_1} e^{2\theta_0 R_\infty}.$$

This can be proved by using the fact that

$$P_{\theta_1}(R_\infty \in dy) = \frac{P_{\theta_1}(S_{\tau_+} > y)}{E_{\theta_1} S_{\tau_+}} dy.$$

Further, we note that

$$E_{\theta_0} S_N = E_{\theta_0}[S_N; S_N < 0] + E_{\theta_0}[S_N; S_N > d]$$

and

$$
\begin{aligned}
E_{\theta_0}[S_N; S_N < 0] &= E_{\theta_0} S_{\tau_-} - E_{\theta_0}[S_{\tau_-}; S_N > d] \\
&= E_{\theta_0} S_{\tau_-} - E_{\theta_0}[E_{\theta_0}[S_{\tau_-}|S_N]; S_N > d] \\
&= E_{\theta_0} S_{\tau_-} - E_{\theta_1} R_\infty P_{\theta_0}(S_N > d)\left(1 + o(e^{-rd})\right).
\end{aligned}
$$

By combining the above results, we have

$$
\begin{aligned}
ARL_0 &= \frac{E_{\theta_0} S_N}{-\theta_0 P_{\theta_0}(S_N > d)} \\
&= \frac{1}{-\theta_0}[\frac{1}{-2\theta_0}\left(e^{-2\theta_0 d}\left(E_{\theta_1} e^{2\theta_0 R_\infty}\right)^{-2} - 1\right) \\
&\quad - \left(E_{\theta_1} R_\infty + E_{\theta_0}(S_N|S_N > d)\right)]\left(1 + o(e^{-rd})\right).
\end{aligned}
$$

By using the same technique as in Lemma 1.4, we have

Lemma 1.6:
$$E_{\theta_0}[S_N - d; S_N > d] = e^{2\theta_0 d} E_{\theta_1}\left(R_\infty e^{2\theta_0 R_\infty}\right) P_{\theta_0}(S_N > d)\left(1 + o(e^{-rd})\right).$$

Finally, we get

Theorem 1.1: *As $d \to \infty$,*

$$
\begin{aligned}
ARL_0 &= \frac{1}{(2\theta_0)^2/2}[e^{-2\theta_0 d}\left(E_{\theta_1} e^{2\theta_0 R_\infty}\right)^{-2} - 1 + 2\theta_0(d + E_{\theta_1} R_\infty \\
&\quad + \left(E_{\theta_1} e^{2\theta_0 R_\infty}\right)^{-1} E_{\theta_1}\left(R_\infty e^{2\theta_0 R_\infty}\right))]\left(1 + o(e^{-rd})\right).
\end{aligned}
$$

We consider the local behavior by further letting $\theta_0 \to 0$.
The Taylor expansion gives

$$E_{\theta_1} e^{2\theta_0 R_\infty} = 1 + 2\theta_0 E_{\theta_1} R_\infty + \frac{(2\theta_0)^2}{2} E_{\theta_1} R_\infty^2 + o\left(\theta_0^2\right).$$

A result from Corollary 1.3 gives

$$E_{\theta_1} R_\infty = \rho - \theta_0(\rho_2 - \rho^2) + o(\theta_0),$$

where

$$\rho_2 = E_0 R_\infty^2.$$

Thus,

$$E_{\theta_1} e^{2\theta_0 R_\infty} = e^{2\theta_0 \rho} \left(1 + o(\theta_0^2)\right).$$

On the other hand,

$$
\begin{aligned}
\frac{E_{\theta_1} \left(R_\infty e^{2\theta_0 R_\infty}\right)}{E_{\theta_1} e^{2\theta_0 R_\infty}} &= E_{\theta_1} R_\infty + 2\theta_0 \left(E_{\theta_1} R_\infty^2 - (E_{\theta_1} R_\infty)^2\right) + o(\theta_0) \\
&= \rho + \theta_0(\rho_2 - \rho^2) + o(\theta_0).
\end{aligned}
$$

Combining these approximations, we have

$$E_{\theta_1} R_\infty + \left(E_{\theta_1} e^{2\theta_0 R_\infty}\right)^{-1} E_{\theta_1} \left(R_\infty e^{2\theta_0 R_\infty}\right) = 2\rho + o(\theta_0).$$

The approximated formula for ARL_0 is thus obtained, and the similar technique can be used to derive the one for ARL_1.

Table 1.1 gives some comparisons between the approximated values with the numerical values given in van Dobben de Bruyn (1968). The table is reproduced from Wu (1994).

Table 1.1: Average Run Lengths

d	θ_1	ARL_0		ARL_1	
		Num.	Approx.	Num.	Approx
2.0	0.0	10.0	10.2	10.0	10.2
	0.2	15.9	16.02	6.86	6.85
	0.4	28.0	28.30	5.06	5.04
	0.6	54.0	55.37	3.96	3.92
2.5	0.0	13.4	13.44	13.4	13.44
	0.2	23.3	23.34	8.73	8.71
	0.4	46.1	46.40	6.24	6.21
	0.6	104.0	105.54	4.79	4.74
3.0	0.0	17.3	17.36	17.3	17.36
	0.2	32.8	32.83	10.7	10.69
	0.4	73.6	74.01	7.44	7.40
4.0	0.0	26.6	26.69	26.6	26.69
	0.2	60.3	60.37	14.9	14.91
	0.4	178.0	178.81	9.88	9.84
	0.6	660.0	673.81	7.28	7.22
5.0	0.0	38.1	38.02	38.1	38.02
	0.2	104.0	103.92	19.4	19.39
	0.4	414.0	415.11	12.4	12.31
6.0	0.0	51.6	51.35	51.6	51.35
	0.2	171.0	171.34	24.0	24.04
	0.4	940.0	944.06	14.9	14.80

2
Change-Point Estimation

2.1   Asymptotic Quasistationary Bias

For notational convenience, we denote by $\{S'_n\}$, for $n \geq 0$, an independent copy of $\{S_n\}$ and

$$M = \sup_{0 \leq k < \inf} S'_k,$$

as the maximum of $\{S'_n\}$ when $S'_0 = 0$ and

$$\sigma_M = \arg \sup_{0 \leq k < \infty} S'_k$$

as the corresponding maximum point.

Conditional on $N > \nu$, depending on whether $\hat{\nu} > \nu$ or $\hat{\nu} < \nu$, we can write the bias as

$$\hat{\nu} - \nu = (\hat{\nu} - \nu)I_{[\hat{\nu} > \nu]} - (\nu - \hat{\nu})I_{[\hat{\nu} < \nu]},$$

where I_A denotes the indicator function for the event A. The notations are consistent with Chapter 1 and, in addition, we shall denote by $E_{\theta_0 \theta}[.]$ the expectation when both P_{θ_0} and P_{θ} are involved.

Pollak and Siegmund (1986) shows that the quasistationary distribution of T_ν converges to the stationary distribution of T_ν as $d \to \infty$. That means,

$$\lim_{d\to\infty} \lim_{\nu\to\infty} P_{\theta_0}(T_\nu < y | N > \nu) = \lim_{d\to\infty} \lim_{\nu\to\infty} P_{\theta_0}(T_\nu < y)$$
$$= P_{\theta_0}(M < y),$$

as we note that T_ν is asymptotically equivalent to M in distribution as $\nu \to \infty$.

This implies that when the change occurs, T_ν is asymptotically distributed as M.

Thus, the event $\{\hat{\nu} > \nu\}$ is asymptotically equivalent to the event $\{\tau_{-M} < \infty\}$, i.e., the random walk S_n eventually goes below to zero with initial starting point M. Given $\hat{\nu} > \nu$, the bias $\hat{\nu} - \nu$ is asymptotically equal to τ_{-M} plus the length, say γ_m for a CUSUM process T_n starting from zero until the last zero point time under $P_\theta(.)$. Define

$$E[X; A] = E[X I_A].$$

As $d, \nu \to \infty$, we have

$$E^\nu[\hat{\nu} - \nu; \hat{\nu} > \nu] \rightarrow E_{\theta_0 \theta}[\tau_{-M} + \gamma_m; \tau_{-M} < \infty]$$
$$= E_{\theta_0 \theta}[\tau_{-M}; \tau_{-M} < \infty] + E_\theta[\gamma_m] P_{\theta_0 \theta}(\tau_{-M} < \infty).$$

We can see that γ_m is a geometric summation of i.i.d. random variables distributed as $\{\tau_-; \tau_- < \infty\}$ with terminating probability $P_\theta(\tau_- = \infty)$. Thus, we have

Lemma 2.1:

$$E_\theta \gamma_m = \frac{E_\theta[\tau_-; \tau_- < \infty]}{P_\theta(\tau_- = \infty)}.$$

On the other hand, given $\hat{\nu} < \nu$, by looking at T_k backward in time starting from ν, we see that $T_{\nu-k}$ behaves like a random walk $\{S'_k\}$ for $k \geq 0$ with maximum value M and thus, $\hat{\nu} - \nu$ is asymptotically distributed as the maximum point σ_M. Thus, as $d, \nu \to \infty$, we have

$$E^\nu[\nu - \hat{\nu}; \hat{\nu} < \nu] \rightarrow E_{\theta_0 \theta}[\sigma_M; \tau_{-M} = \infty]$$
$$= E_{\theta_0}[\sigma_M P_\theta(\tau_{-M} = \infty)].$$

A similar argument is used in Srivastava and Wu (1999) for the continuous-time analog.

Summarizing the results, we get the following asymptotic first-order result.

Theorem 2.1 : *As $\nu, d \to \infty$,*

$$E^\nu[\hat{\nu} - \nu | N > \nu] \rightarrow E_{\theta_0 \theta}[\tau_{-M}; \tau_{-M} < \infty] + P_{\theta_0 \theta}(\tau_{-M} < \infty)\frac{E_\theta[\tau_-; \tau_- < \infty]}{P_\theta(\tau_- = \infty)}$$
$$-E_{\theta_0 \theta}[\sigma_M; \tau_{-M} = \infty],$$

$$E^\nu[|\hat{\nu} - \nu| | N > \nu] \rightarrow E_{\theta_0 \theta}[\tau_{-M}; \tau_{-M} < \infty] + P_{\theta_0 \theta}(\tau_{-M} < \infty)\frac{E_\theta[\tau_-; \tau_- < \infty]}{P_\theta(\tau_- = \infty)}$$
$$+E_{\theta_0 \theta}[\sigma_M; \tau_{-M} = \infty].$$

2.2 Second-Order Approximation

In this section, we shall derive the second-order approximation for the asymptotic bias given in Theorem 2.1 in order to investigate the bias numerically and also see some local properties by further assuming θ_0 and θ approach zero at the same order. The main theoretical tool is the strong renewal theorem given in Section 1.3 and its applications to ladder variables.

There are five terms in Theorem 2.1 which will be evaluated in a sequence of lemmas. Most of the results generalizes the ones given in Wu (1999) in the fixed sample size case with normal distribution. However, the technique used here is much more general and can be used for any distribution of exponential family type and also raises more difficulties due to sequential sampling plan.

The first lemma gives the approximation for the expected length τ_- if the random walk goes below zero.

Lemma 2.2: *As $\theta \to 0$,*

$$E_\theta[\tau_-; \tau_- < \infty] = \frac{E_0 S_{\tau_-}}{\tilde{\mu}} \exp\left(\theta\rho_- + \frac{\theta^2}{2}\left(\rho_-^{(2)} - \rho_-^2 - \frac{5\beta_1}{E_0 S_{\tau_-}}\right)\right)(1 + o(\theta^2)),$$

where β_1 is given in Lemma 1.1.

Proof: By using Wald's likelihood ratio identity by changing the measure $P_\theta(.)$ to $P_{\tilde{\theta}}(.)$ and Lemma 1.1, we have

$$
\begin{aligned}
E_\theta[\tau_-; \tau_- < \infty] &= E_{\tilde{\theta}}[\tau_- e^{\Delta S_{\tau_-}}] \\
&= \frac{1}{\tilde{\mu}} E_{\tilde{\theta}} S_{\tau_-} + \Delta E_{\tilde{\theta}}(\tau_- S_{\tau_-}) + \frac{\Delta^2}{2} E_{\tilde{\theta}}\left(\tau_- S_{\tau_-}^2\right) + o(\Delta).
\end{aligned}
$$

After some algebraic simplifications, we get the result.

Corollary 2.1: *As $\theta \to 0$,*

$$E_\theta \gamma_m = -\frac{1}{\Delta\tilde{\mu}} e^{-(2\beta_1/E_0 S_{\tau_-})\theta^2}(1 + o(\theta^2)).$$

To evaluate $P_{\theta_0\theta}(\tau_{-M} < \infty)$, we follow a similar technique used in Wu (1999) and only the main steps are provided.

First, by conditioning on whether $M = 0$ or $M > 0$, we have

$$
\begin{aligned}
P_{\theta_0\theta}(\tau_{-M} < \infty) &= P_\theta(\tau_- < \infty)P_{\theta_0}(\tau_+ = \infty) \\
&\quad + P_{\theta_0\theta}(\tau_{-M} < \infty; M > 0). \quad (2.1)
\end{aligned}
$$

From Lemma 1.2, we have

$$
\begin{aligned}
P_\theta(\tau_- < \infty)P_{\theta_0}(\tau_+ = \infty) &= \Delta_0 E_0 S_{\tau_+} e^{\theta_0\rho_+}(1 + \Delta E_0 S_{\tau_+}) + o(\theta^2) \\
&= \Delta_0 E_0 S_{\tau_+} e^{\theta_0 S_{\tau_+}} - \frac{\Delta\Delta_0}{2} + o(\theta^2).
\end{aligned}
$$

For the second term in (2.1), by using the Wald's likelihood ratio identity by changing parameters θ to $\tilde{\theta}$ and θ_0 to θ_1, we have

$$
\begin{aligned}
P_\theta(\tau_x < \infty) &= E_{\tilde{\theta}} e^{\Delta S_{\tau-x}} \\
&= e^{-\Delta x} E_{\tilde{\theta}} e^{\Delta R_{-x}},
\end{aligned}
$$

and

$$
\begin{aligned}
P_{\theta_0}(M > x) &= P_{\theta_0}(\tau_x < \infty) \\
&= e^{-\Delta_0 x} E_{\theta_1} e^{\Delta_0 R_x}.
\end{aligned}
$$

From Corollary 1.1, we know

$$
E_0 R_x - \rho_+ = O\left(e^{-rx}\right) \quad \text{and} \quad E_0 R_{-x} - \rho_- = O\left(e^{-rx}\right),
$$

as $x \to \infty$. Now, we write

$$
P_{\theta_0 \theta}(\tau_{-M} < \infty, M > 0)
$$

$$
\begin{aligned}
&= -\int_0^\infty P_\theta(\tau_{-x} < \infty) dP_{\theta_0}(M > x) \\
&= -\int_0^\infty E_{\tilde{\theta}} e^{\Delta S_{\tau-x}} dE_{\theta_1} e^{-\Delta_0 S_{\tau x}} \\
&= -\int_0^\infty e^{-\Delta(x-\rho_-)} de^{-\Delta_0(x+\rho_+)} \\
&\quad -\int_0^\infty e^{-\Delta(x-\rho_-)} d\left(e^{-\Delta_0(x+\rho_+)}\left(E_{\theta_1} e^{-\Delta_0(R_x-\rho_+)}\right) - 1\right) \\
&\quad -\int_0^\infty e^{-\Delta(x-\rho_-)}\left(E_{\theta_0} e^{\Delta(R_{-x}-\rho_-)} - 1\right) de^{-\Delta_0(x+\rho_+)} \\
&\quad -\int_0^\infty e^{-\Delta(x-\rho_-)}\left(E_{\theta_0} e^{\Delta(R_{-x}-\rho_-)} - 1\right) \\
&\quad \times d\left(e^{-\Delta_0(x+\rho_+)}\left(E_{\theta_1} e^{-\Delta_0(R_x-\rho_+)}\right) - 1\right).
\end{aligned} \tag{2.2}
$$

The first term in (2.2) is

$$
\frac{\Delta_0}{\Delta + \Delta_0} e^{\Delta \rho_- - \Delta_0 \rho_+}.
$$

The third term in (2.2) is approximately equal to

$$
\Delta \Delta_0 \int_0^\infty (E_0 R_{-x} - \rho_-) dx + o(\theta^2).
$$

The fourth term in (2.2) is

$$
\Delta \Delta_0 \int_0^\infty E_0(R_{-x} - \rho_-) dE_0(R_x - \rho_+) + o(\theta^2).
$$

The second term in (2.2), by integrating by parts, can be approximated as

$$e^{\Delta \rho_-}\left(P_{\theta_0}(\tau_+ < \infty) - e^{-\Delta_0 \rho_+}\right) + \Delta \Delta_0 \int_0^\infty (E_0 R_x - \rho_+)dx + o(\theta^2).$$

Combining the above approximations, we have

Lemma 2.3: *As $\theta_0, \theta \to 0$ at the same order,*

$$
\begin{aligned}
P_{\theta_0 \theta}(\tau_{-M} < \infty) \;=\; & \frac{\Delta_0}{\Delta + \Delta_0} e^{\Delta \rho_- - \Delta_0 \rho_+} + e^{\Delta \rho_-}\left(1 - e^{-\Delta_0 \rho_+}\right) \\
& + \Delta \Delta_0 \left(-\frac{1}{2} - \rho_- E_0 S_{\tau_-} + \int_0^\infty (E_0 R_{-x} - \rho_-)dx\right. \\
& + \int_0^\infty (E_0 R_{-x} - \rho_-)d(E_0 R_x - \rho_+) + \int_0^\infty (E_0 R_x - \rho_+)dx\bigg) \\
& + o(\theta^2).
\end{aligned}
$$

The evaluation of $E_{\theta_0 \theta}[\tau_{-M}; \tau_{-M} < \infty]$ is similar.

Lemma 2.4: *As $\theta_0, \theta \to 0$ at the same order,*

$$
\begin{aligned}
E_{\theta_0 \theta}[\tau_{-M}; \tau_{-M} < \infty] \;=\; & -\frac{\Delta_0}{\tilde{\mu}}\left(\frac{1}{(\Delta + \Delta_0)^2} - \frac{\rho_-}{\Delta + \Delta_0}\right) \\
& -\frac{\Delta_0}{\tilde{\mu}}\left(\frac{1}{2} + \rho_-(E_0 S_{\tau_+} - \rho_+) - \int_0^\infty (E_0 R_{-x} - \rho_-)dx\right. \\
& -\int_0^\infty (E_0 R_{-x} - \rho_-)d(E_0 R_x - \rho_+) \\
& -\int_0^\infty (E_0 R_x - \rho_+)dx\bigg) + o(1).
\end{aligned}
$$

Proof: Again depending on whether $\{M = 0\}$ or $\{M > 0\}$, we have

$$E_{\theta_0 \theta}[\tau_{-M}; \tau_{-M} < \infty]$$

$$= E_\theta[\tau_-; \tau_- < \infty]P_{\theta_0}(\tau_+ = \infty) - \int_0^\infty E_\theta[\tau_{-x}; \tau_{-x} < \infty]dP_{\theta_0}(\tau_x < \infty). \quad (2.3)$$

The first term in (2.3) can be approximated by using Lemmas 2.2 and 1.2 as

$$
\begin{aligned}
E_\theta[\tau_-; \tau_- < \infty]P_{\theta_0}(\tau_+ = \infty) \;=\; & \frac{\mu_0}{\tilde{\mu}} + o(1) \\
=\; & -\frac{\Delta}{2\tilde{\mu}} + o(1).
\end{aligned}
$$

For the second term in (2.3), by conditioning on the value of M given $M > 0$ we use the similar techniques as in Lemma 2.3 and write it as

$$-\int_0^\infty E_\theta[\tau_{-x}; \tau_{-x} < \infty]dP_{\theta_0}(\tau_x < \infty)$$

$$= -\int_0^\infty E_{\tilde{\theta}}\left[\tau_{-x}e^{-\Delta(x-R_{-x})}\right]dE_{\theta_1}e^{-\Delta_0(x+R_x)}$$

$$= -\int_0^\infty E_{\tilde{\theta}}(\tau_{-x})e^{-\Delta(x-\rho_-)}dE_{\theta_1}e^{-\Delta_0(x+R_x)} + o(1)$$

$$= -\frac{1}{\tilde{\mu}}[\int_0^\infty (-x + E_0 R_{-x})e^{-\Delta(x-\rho_-)}de^{-\Delta_0(x+\rho_+)}$$

$$- \Delta_0 \int_0^\infty (-x + E_0 R_{-x})d(E_0 R_x - \rho_+)] + o(1). \tag{2.4}$$

The first term in (2.4) is approximately equal to

$$-\frac{\Delta_0}{\tilde{\mu}}\left[e^{\Delta\rho_- - \Delta_0\rho_+}\left(\frac{1}{(\Delta+\Delta_0)^2} - \frac{\rho_-}{\Delta+\Delta_0}\right) - \int_0^\infty (E_0 R_{-x} - \rho_-')dx\right] + o(1).$$

The second term in (2.4) is equal to

$$-\frac{\Delta_0}{\tilde{\mu}}\left[\int_0^\infty (x - \rho_-)d(E_0 R_x - \rho_+) - \int_0^\infty (E_0 R_{-x} - \rho_-)d(E_0 R - x - \rho_+)\right]$$

$$= -\frac{\Delta_0}{\tilde{\mu}}[\rho_-(E_0 S_{\tau_+} - \rho_+) - \int_0^\infty (E_0 R - x - \rho_+)dx$$

$$- \int_0^\infty (E_0 R_{-x} - \rho_-)d(E_0 R_x - \rho_+)].$$

Combining the above results, we complete the proof.

Finally, we evaluate $E_{\theta_0\theta}[\sigma_M; \tau_{-M} < \infty]$. We first write

$$E_{\theta_0\theta}[\sigma_M; \tau_{-M} < \infty] = E_{\theta_0}\sigma_M - E_{\theta_0}\left[\sigma_M E_{\tilde{\theta}}e^{\Delta(-M+R_{-M})}\right]. \tag{2.5}$$

For the second term in (2.5), we write

$$E_{\theta_0}\left[\sigma_M E_{\tilde{\theta}}e^{\Delta(-M+R_{-M})}\right] = E_{\theta_0}\left[\sigma_M e^{-\Delta M}\right]e^{-\Delta\rho_-}$$

$$+ E_{\theta_0}\left[\sigma_M e^{-\Delta M}\left(E_{\tilde{\theta}}e^{\Delta R_{-M}} - e^{\Delta\rho_-}\right)\right].$$

To evaluate $E_{\theta_0}[\sigma_M e^{-\Delta M}]$, we note that under $P_{\theta_0}(.)$

$$(\sigma_M, M) =^d (\tau_+^{(K)}, S_{\tau_+^{(K)}}),$$

where $=^d$ means equivalence in distribution, $\tau_+^{(k)}$ is the kth ladder epoch defined in Section 1.3, and

$$K = \inf\{k > 0: \ \tau_+^{(k)} < \infty\}.$$

Note that K is a geometric random variable with

$$P(K = k) = p^k(1 - p)$$

for $k \geq 0$, with terminating probability

$$1 - p = P_{\theta_0}(\tau_+ = \infty).$$

For given $K = k$, $(\tau_+^{(k)}, S_{\tau_+^{(k)}})$ is, in distribution, equivalent to the sum of k i.i.d. random variables distributed as (τ_+, S_{τ_+}).

Thus,

$$
\begin{aligned}
E_{\theta_0}\left[\sigma_M e^{-\Delta M}\right] &= E_{\theta_0}\left[\tau_+^{(K)} \exp(-\Delta S_{\tau_+^{(K)}})\right] \\
&= \sum_{k=1}^{\infty} E_{\theta_0}\left[\tau_+^{(k)} \exp(-\Delta S_{\tau_+^{(k)}}); K = k\right] \\
&= \sum_{k=1}^{\infty} k E_{\theta_0}\left[\tau_+ \exp(-\Delta S_{\tau_+}); \tau_+ < \infty\right] \\
&\quad \times \left(E_{\theta_0}\left[\exp(-\Delta S_{\tau_+}); \tau_+ < \infty\right]\right)^{k-1} P_{\theta_0}(\tau_+ = \infty) \\
&= \frac{E_{\theta_0}\left[\tau_+ \exp(-\Delta S_{\tau_+}); \tau_+ < \infty\right]}{\left(1 - E_{\theta_0}\left[\exp(-\Delta S_{\tau_+}); \tau_+ < \infty\right]\right)^2} P_{\theta_0}(\tau_+ = \infty).
\end{aligned}
$$

The next two lemmas give the approximations for the related quantities.

Lemma 2.5: *As $\theta_0, \theta \to$ at the same order,*

$$1 - E_{\theta_0}\left[\exp(-\Delta S_{\tau_+}); \tau_+ < \infty\right]$$

$$= (\Delta + \Delta_0) E_0 S_{\tau_+} \exp\left(-(\Delta - \theta_0)\rho_+ + \frac{1}{2}(\Delta - \theta_0)^2(\rho_+^{(2)} - \rho_+^2) - \frac{\theta_0^2}{2}\frac{\alpha_1}{E_0 S_{\tau_+}}\right)$$

$$\times (1 + o(\theta^2)).$$

Proof: Using Wald's likelihood ratio identity, we have

$$1 - E_{\theta_0}\left[\exp(-\Delta S_{\tau_+}); \tau_+ < \infty\right] = 1 - E_{\theta_1} e^{-(\Delta + \Delta_0) S_{\tau_+}}.$$

The Taylor series expansion following the lines of Lemma 1.2 will give the result after some algebraic simplification.

In particular, by letting $\Delta = 0$, we have

$$P_{\theta_0}(\tau_+ = \infty) = \Delta_0 E_0 S_{\tau_+} \exp\left(\theta_0 \rho_+ + \frac{1}{2}\theta_0^2\left(\rho_+^{(2)} - \rho_+^2 - \frac{\alpha}{E_0 S_{\tau_+}}\right)\right)(1 + o(\theta_0^2)).$$

The following lemma can be proved similarly as for Lemma 2.2, and its proof is omitted.

Lemma 2.6: *As $\theta_0, \theta \to$ at the same order,*

$$E_{\theta_0}\left[\tau_+ e^{-\Delta S_{\tau_+}}; \tau_+ < \infty\right] = \frac{E_0 S_{\tau_+}}{\mu_1} \exp(-(\Delta - \theta_0)\rho_+ + \frac{1}{2}(\Delta - \theta_0)^2(\rho_+^{(2)} - \rho_+^2)$$

$$- \frac{\theta_1^2}{2} \frac{\alpha_1}{E_0 S_{\tau_+}} - \theta_1(\Delta + \Delta_0)\frac{\alpha_1}{E_0 S_{\tau_+}})(1 + o(\theta^2)).$$

In particular, by letting $\Delta = 0$, we have

$$E_{\theta_0}[\tau_+; \tau_+ < \infty] = \frac{E_0 S_{\tau_+}}{\mu_1}\exp\left(\theta_0\rho_+ + \frac{\theta_0^2}{2}\left(\rho_+^{(2)} - \rho_+^2 - \frac{5\alpha_1}{E_0 S_{\tau_+}}\right)\right)(1 + o(\theta^2)).$$

On the other hand, by conditioning on the value of M, we have

$$E_{\theta_0}\left[\sigma_M e^{-\Delta M}\left(E_{\tilde{\theta}} e^{\Delta R_{-M}} - e^{\Delta\rho_-}\right)\right] \tag{2.6}$$

$$= \Delta E_{\theta_0}[\sigma_M (E_0 R_{-M} - \rho_-)](1 + o(1))$$

$$= -\Delta \int_0^\infty E_{\theta_0}[\sigma_x | M = x](E_0 R_{-x} - \rho_-)dP_{\theta_0}(M > x)$$

$$= \Delta\Delta_0 \int_0^\infty E_{\theta_0}[\sigma_x | M = x](E_0 R_{-x} - \rho_-)d(x + E_0 R_x)(1 + o(1)).$$

Since

$$E_{\theta_0}[S_{\tau_+}; \tau_+ < \infty] = E_0 S_{\tau_+}(1 + o(1)),$$

as $\theta_0 \to 0$, thus, $K = O_p(x)$, where $O_p(.)$ means at the same order in probability. This implies

$$E_{\theta_0}[\sigma_x | M = x] = O(\frac{x}{\mu_0}).$$

Thus, (2.6) is at the the order of $O(\theta)$.

By letting $\Delta = 0$, the first term of (2.5) can be evaluated by combining Lemmas 2.5 and 2.6.

Lemma 2.7: *As $\theta_0 \to 0$,*

$$E_{\theta_0}\sigma_M = \frac{E_{\theta_0}[\tau_+; \tau_+ < \infty]}{P_{\theta_0}(\tau_+ = \infty)}$$

$$= \frac{1}{\Delta_0\mu_1}e^{-\theta_0^2(2\alpha_1/E_0 S_{\tau_+})}(1 + o(\theta_0^2)).$$

Finally, we have the following result:

Lemma 2.8: *As $\theta_0, \theta \to 0$,*

$$E_{\theta_0\theta}[\sigma_M; \tau_{-M} = \infty] = \frac{1}{\Delta_0\mu_1}e^{-\theta_0^2(2\alpha_1/E_0 S_{\tau_+})}$$

$$-\frac{\Delta_0}{\mu_1(\Delta+\Delta_0)^2}e^{\gamma\Delta/3-2\theta(\theta-\theta_0)(\rho_+^{(2)}-\rho_+^2-\alpha_1/E_0S_{\tau_+})-2(\theta-\theta_0)^2(\alpha_1/E_0S_{\tau_+})}(1+o(\theta^2)).$$

Combining Lemmas 2.1 − 2.8 and Lemmas 1.1 − 1.2, we have the following local second-order expansion for the asymptotic bias of $\hat{\nu}$.

Theorem 2.2: *As $\theta_0, \theta \to 0$ at the same order, we have*

$$\lim_{d\to\infty}\lim_{\nu\to\infty}E^\nu[\hat{\nu}-\nu|N>\nu]=-\frac{1}{\tilde{\mu}\Delta}\left(\frac{\Delta}{\Delta+\Delta_0}e^{\Delta\rho_--\Delta_0\rho_+}+e^{\Delta\rho_-}\left(1-e^{-\Delta_0\rho_+}\right)\right)$$

$$-\frac{\Delta}{\tilde{\mu}}\left(\frac{1}{(\Delta+\Delta_0)^2}-\frac{\rho_-}{\Delta+\Delta_0}\right)e^{\Delta\rho_--\Delta_0\rho_+}$$

$$+\frac{\theta_0}{\theta-\theta_0}\frac{\beta_1}{E_0S_{\tau_+}}+\frac{2\theta_0}{\theta}\rho_-\rho_+$$

$$-\frac{\theta}{\theta-\theta_0}\left(\rho_+^{(2)}-\rho_+^2-\frac{\alpha_1}{E_0S_{\tau_+}}\right)+o(1).$$

A similar result can be obtained for the absolute bias $E^\nu[|\hat{\nu}-\nu||N>\nu]$. In particular, when $\theta=\theta_1$, we have the following result.

Corollary 2.2: *As $\theta=\theta_1\to 0$,*

$$\lim_{d\to\infty}\lim_{\nu\to\infty}E^\nu[\hat{\nu}-\nu|N>\nu]$$

$$=-\frac{3}{4\Delta_0}\left(\frac{1}{\mu_0}+\frac{1}{\mu_1}\right)-\frac{\gamma}{12}\left(\frac{1}{\mu_0}+\frac{1}{\mu}\right)-\frac{1}{2}\frac{\beta_1}{E_0S_{\tau_-}}$$

$$-\frac{1}{2}\left(\rho_+^{(2)}-\rho_+^2-\frac{\alpha_1}{E_0S_{\tau_+}}\right)+o(1)$$

$$=-\frac{\gamma}{4\theta_0}+\frac{17\gamma^2}{288}-\frac{\kappa}{16}-\frac{1}{2}\frac{\beta_1}{E_0S_{\tau_-}}-\frac{1}{2}\left(\rho_+^{(2)}-\rho_+^2-\frac{\alpha_1}{E_0S_{\tau_+}}\right)+o(1).$$

Proof: The first equation is a direct simplification of Theorem 2.2. For the second equation, we note that as $\theta_0\to 0$,

$$\mu_1 = \theta_1e^{\theta_1(\gamma/2)+\theta_1^2(\kappa/6-\gamma^2/8)}(1+o(\theta_1^2)),$$

$$\Delta_0 = 2\theta_1e^{\theta_1(\gamma/6)+\theta_1^2(\gamma^2/24)}(1+o(\theta_1^2)),$$

$$\theta_0 = -\theta_1e^{\theta_1(\gamma/3)+\theta_1^2(\gamma^2/18)}(1+o(\theta_1^2)),$$

$$\mu_0 = -\theta_1e^{-\theta_1(\gamma/6)+\theta_1^2(\kappa/6-17\gamma^2/72)}(1+o(\theta_1^2)).$$

Some tedious simplifications give the expected results.

Therefore, the local bias of $\hat{\nu}$ is largely affected by the skewness γ. If $\gamma > 0$, the local bias becomes positive. If $F_0(x)$ is symmetric, from Corollary 1.2, we have

$$\rho_+^{(2)} - \rho_+^2 - \frac{\alpha_1}{E_0 S_{\tau_+}} = \frac{\kappa}{6},$$

and thus,

$$E^\nu[\hat{\nu} - \nu | N > \nu] \approx -\frac{7}{48}\kappa - \frac{1}{2}\frac{\beta_1}{E_0 S_{\tau_-}} + o(1),$$

which, surprisingly, is a nonzero constant, in contrast to the fixed sample size case as given in the normal case of the next section.

2.3   Two Examples

In this section, we present two cases: normal and exponential distributions.

2.3.1   Normal Distribution

From Example 1.1, the approximations for the related quantities are simplified as

$$
\begin{aligned}
E_\theta(\gamma_m) &= \frac{1}{2\theta^2} - \frac{1}{4} + o(1), \\
P_{\theta_0\theta}(\tau_{-M} < \infty) &= -\frac{\theta_0}{\theta - \theta_0}e^{-\theta(\theta-\theta_0)} + o(\theta^2), \\
E_{\theta_0\theta}[\tau_{-M}; \tau_{-M} < \infty] &= -\frac{\theta_0}{2\theta(\theta - \theta_0)^2}e^{(\theta-\theta_0)^2} + o(1), \\
E_{\theta_0\theta}[\sigma_M; \tau_{-M} < \infty] &= \frac{1}{2\theta_0^2} - \frac{1}{2(\theta - \theta_0)^2} + o(1).
\end{aligned}
$$

Summarizing the above results, we have the following corollary:

Corollary 2.3: *As $\theta_0, \theta \to 0$ at the same order,*

$$\lim_{d\to\infty}\lim_{\nu\to\infty} E^\nu[\hat{\nu} - \nu | N > \nu] = \frac{1}{2\theta^2} - \frac{1}{2\theta_0^2} + \frac{\theta_0}{4(\theta - \theta_0)} + o(1),$$

$$\lim_{d\to\infty}\lim_{\nu\to\infty} E^\nu[|\hat{\nu} - \nu| | N > \nu] = \frac{1}{2}\left(\frac{1}{\theta^2} + \frac{1}{\theta_0^2} - \frac{2}{(\theta - \theta_0)^2}\right) + \frac{\theta_0}{4(\theta - \theta_0)} + o(1).$$

At $\theta = \theta_1 = -\theta_0$,

$$\lim_{d\to\infty}\lim_{\nu\to\infty} E^\nu[\hat{\nu} - \nu | N > \nu] = -\frac{1}{8} + o(1),$$

$$\lim_{d\to\infty}\lim_{\nu\to\infty} E^\nu[|\hat{\nu} - \nu| | N > \nu] = \frac{3}{4\theta_0^2} - \frac{1}{8} + o(1).$$

Remark: Wu (1999) considered the bias of the estimator in the large fixed sample size case, which corresponds to the maximum point of a two-sided random walk, and obtained the following result:

$$E^\nu[\hat{\nu} - \nu] = \frac{1}{2\theta^2} - \frac{1}{2\theta_0^2} + \frac{1}{4}\frac{\theta + \theta_0}{\theta - \theta_0} + o(1);$$

and at $\theta = -\theta_0$,

$$E^\nu[|\hat{\nu} - \nu|] = \frac{3}{4\theta_0^2} - \frac{1}{4} + o(1).$$

Srivastava and Wu (1999) also considered the continuous-time analog in the sequential sampling plan case, which gives

$$\lim_{d\to\infty} \lim_{\nu\to\infty} E^\nu[\hat{\nu} - \nu | N > \nu] = \frac{1}{2\theta^2} - \frac{1}{2\theta_0^2},$$

which is zero at $\theta = -\theta_0$ and at $\theta = -\theta_0$,

$$\lim_{d\to\infty} \lim_{\nu\to\infty} E^\nu[\hat{\nu} - \nu | N > \nu] = \frac{3}{4\theta_0^2}.$$

We see that the sequential sampling plan has local effect at the second-order in the discrete-time case and is negative at $\theta = -\theta_0$.

To show the accuracy of the second-order approximations, we conduct a simple simulation study. For $d = 10$ and $\theta_0 = -0.25, -0.5$, we let $\nu = 50$ and 100. One thousand replications of the CUSUM charts are simulated for several values for θ. Only those runs with $N > \nu$ are used for calculating $\hat{\nu}$. Table 2.1 gives the comparison between the simulated results and approximated values. The approximated values from Corollary 2.4 are given in the parentheses. We see that the approximations are generally good. The case $\nu = 100$ shows quite satisfactory results. Also, we see that approximations for the case $\theta_0 = -0.5$ perform better than those for the case $\theta_0 = -0.25$. The reason is that our results are given by first assuming $d, \nu \to \infty$ and then letting $\theta_0, \theta \to 0$. The effect of ν is very little. However, as the local bias is at the order $O(1/\theta_0^2)$ at $\theta = -\theta_0$, which approaches ∞ as $\theta_0 \to 0$, there could be an error term at the order, say $O(1/(d\theta))$ for finitely large d. Thus, the approximation may perform better for $\theta = -\theta_0 = 0.5$. The case when θd approaches a constant, called moderate deviation as considered in Chang (1992), is definitely worths a future study.

2.3.2   Exponential Distribution

Here, we are interested in quick detection of increment in the mean of an exponential distribution from the initial mean 1.

From Example 1.2 and Theorem 2.2, we have the following result:

Corollary 2.4. *As $\theta, \theta_0 \to 0$ at the same order,*

Table 2.1: Biases in the Normal Case

ν	θ_0	θ	$E[\hat\nu - \nu | N > \nu]$	$E[|\hat\nu - \nu| | N > \nu]$
50	-0.25	0.25	$0.113(-0.125)$	$9.737(11.875)$
		0.5	$-4.902(-6.083)$	$7.090(8.139)$
		0.75	$-5.682(-7.174)$	$6.376(7.826)$
		1.0	$-5.768(-7.550)$	$6.188(7.810)$
	-0.5	0.5	$0.268(-0.125)$	$3.02(2.875)$
		0.75	$-1.302(-1.211)$	$2.338(2.149)$
		1.0	$-1.730(-1.583)$	$3.135(1.972)$
100	-0.25	0.25	$1.368(-0.125)$	$11.728(11.875)$
		0.5	$-5.644(-6.083)$	$7.768(8.139)$
		0.75	$-6.181(-7.174)$	$6.942(7.826)$
		1.0	$-6.250(-7.55)$	$6.520(7.810)$
	-0.5	0.5	$-0.223(-0.125)$	$3.052(2.875)$
		0.75	$-1.109(-1.211)$	$2.208(2.149)$
		1.0	$-1.564(-1.583)$	$2.084(1.972)$

$\lim_{d\to\infty} \lim_{\nu\to\infty} E^\nu[\hat\nu - \nu | N > \nu]$

$$
\begin{aligned}
= \; & -\frac{\Delta + 2\Delta_0}{\tilde\mu(\Delta + \Delta_0)^2} e^{-(1/2)\Delta - \Delta_0} \\
& - \frac{1}{\tilde\mu\Delta} e^{-\Delta/2}(1 - e^{-\Delta_0}) - \frac{\Delta_0}{2\tilde\mu(\Delta + \Delta_0)} e^{-(1/2)\Delta - \Delta_0} \\
& - \frac{1}{\Delta_0\mu_1} + \frac{\Delta_0}{\mu_1(\Delta + \Delta_0)^2} e^{(2/3)\Delta} \\
& - \frac{\theta}{\theta - \theta_0} - \frac{\theta_0}{\theta} + \frac{7}{18}\frac{\theta_0}{\theta - \theta_0} + o(1).
\end{aligned}
$$

At $\theta = -\theta_0$,

$$
\lim_{d\to\infty} \lim_{\nu\to\infty} E^\nu[\hat\nu - \nu | N > \nu] = -\frac{1}{2\theta_0} - \frac{17}{24} + o(1).
$$

We see that due to the asymmetry of $F_0(x)$, the local bias becomes positive as θ_0 is small.

2.4   Case Study

In this section, we conduct three classical case studies to illustrate the applications.

Table 2.2: Nile River Flow from 1871 − 1970

Year	Flow	Year	Flow	Year	Flow	Year	Flow
1871	1120	1896	1220	1921	768	1946	1040
1872	1160	1897	1030	1922	845	1947	860
1873	963	1898	1100	1923	864	1948	874
1874	1210	1899	774	1924	862	1949	848
1875	1160	1900	840	1925	698	1950	890
1876	1160	1901	874	1926	845	1951	744
1877	813	1902	694	1927	744	1952	749
1878	1230	1903	940	1928	796	1953	838
1879	1370	1904	833	1929	1040	1954	1050
1880	1140	1905	701	1930	759	1955	918
1881	995	1906	916	1931	781	1956	986
1882	935	1907	692	1932	865	1957	797
1883	1110	1908	1020	1933	845	1958	923
1884	994	1909	1050	1934	944	1959	975
1885	1020	1910	969	1935	984	1960	815
1886	960	1911	831	1936	897	1961	1020
1887	1180	1912	726	1937	822	1962	906
1888	799	1913	456	1938	1010	1963	901
1889	958	1914	824	1939	771	1964	1170
1890	1140	1915	702	1940	676	1965	912
1891	1100	1916	1120	1941	649	1966	746
1892	1210	1917	1100	1942	846	1967	919
1893	1150	1918	832	1943	812	1968	718
1894	1250	1919	764	1944	742	1969	714
1895	1260	1920	821	1945	801	1970	740

2.4.1   Nile River Data (Normal Case):

The following Nile River flow data are reproduced from Cobb(1978), and the data are read in columns.

From a scatterplot, we see there is an obvious change around the year 1900. A qq-normal plot shows the normality is roughly true.

As in Cobb(1978), we assume the independent normality. Assume the pre-change mean is $m_0 = 1100$ and the post-change mean is $m_1 = 850$, with a change magnitude $m_1 - m_0 = 250$ and standard deviation 125.

To apply the CUSUM procedure, we first standardize the data by subtracting all the observations by $m_0 + (m_1 - m_0)/2 = 975$, the average of the pre-change and post-change means, then switch the sign in order to make the change positive, and then divide all the data by 125.

After these transformations, we standardize the observations to x_i for $i =$

1, 2, ..., 100 such that

$$\theta_0 = -1, \quad \theta_1 = 1, \quad and \quad \sigma^2 = 1.$$

Now we form the CUSUM process by letting $T_0 = 0$ and

$$T_i = \max(0, T_{i-1} + x_i),$$

for $i = 1, ..., 100$. The calculated values are reported as follows:

[1] 0.00 0.00 0.10 0.00 0.00 0.00 1.30 0.00 0.00 0.00 0.00 0.32 0.00 0.00 0.00
[16] 0.12 0.00 1.41 1.54 0.22 0.00 0.00 0.00 0.00 0.00 0.00 0.00 0.00 1.61 2.69
[31] 3.50 5.74 6.02 7.16 9.35 9.82 12.09 11.73 11.13 11.18 12.33 14.32 18.47 19.68 21.86
[46] 20.70 19.70 20.85 22.54 23.77 25.42 26.46 27.35 28.26 30.47 31.51 33.36 34.79 34.27 36.00
[61] 37.55 38.43 39.47 39.72 39.65 40.27 41.50 41.22 42.85 45.24 47.85 48.88 50.18 52.05 53.44
[76] 52.92 53.84 54.65 55.66 56.34 58.19 60.00 61.10 60.50 60.95 60.86 62.29 62.70 62.70 63.98
[91] 63.62 64.18 64.77 63.21 63.71 65.54 65.99 68.05 70.14 72.02

The reason we report the CUSUM process instead of drawing graphs is to inspect the path directly rather than guessing from the graph. By looking at the CUSUM process, we see that in the fixed sample size case, the maximum likelihood ratio estimator (the last zero point of the CUSUM process) is $\hat{\nu}_n = 28$ which is the year 1898.

The estimated pre-change mean is the average of the observations from 1970 to 1998, which is -0.982; the estimated post-change mean is 1.00; and the pre-change standard deviation is 1.08 and post-change standard deviation is 1.00. We see that the assumption is roughly correct except for the slight discrepancy from the pre-change standard deviation.

Let us look at the estimator from the sequential sampling plan point of view, which is more natural from the nature of monitoring.

We see that as long as the threshold d is taken larger than 2.00 and less than 70, a change is always signaled and the maximum likelihood estimator $\hat{\nu} = 28$ no matter what the value d is.

Thus, we see that the estimator $\hat{\nu}$ is stable to the selection of the threshold d.

2.4.2   British Coal Mining Disaster (Exponential Case)

The following table gives the intervals in days between successive coal mining disasters in Great Britain for the period 1875−1951. A disaster is defined as involving the death of 10 or more men. The data are taken from Maguire, Pearson, and Wynn(1952) and appeared in many places; the most noticeable is Cox and Lewis (1966). The data are read in columns.

Table 2.3: British Coal Mining Disaster Intervals

378	286	871	66
36	114	48	291
15	108	123	4
31	188	457	369
215	233	498	338
11	28	49	336
137	22	131	19
4	61	182	329
15	78	255	330
72	99	195	312
96	326	224	171
124	275	566	145
50	54	390	75
120	217	72	364
203	113	228	37
176	32	271	19
55	23	208	156
93	151	517	47
59	361	1613	129
315	312	54	1630
59	354	326	29
61	58	1312	217
1	275	348	7
13	78	745	18
189	17	217	1357
345	1205	120	
20	644	275	
81	467	20	

We first explain the transformation on the data in the exponential case in order to fit them into the frame.

Suppose the original data $\{Y_i\}$ follow $\exp(\lambda_0)$ for $i \leq \nu$ and $\exp(\lambda_1)$ for $i > \nu$ where $\lambda_0 > \lambda_1$ are the corresponding hazard rates.

Define

$$\lambda^* = (\lambda_0 - \lambda_1)/\ln(\lambda_0/\lambda_1),$$

and make the following data transformation:

$$X_i = \lambda^* Y_i - 1,$$

for $i = 1, 2, ...$

Denote

$$f_0(x) = e^{-(x+1)},$$

for $x \leq -1$, and then

$$f_\theta(x) = \exp(\theta x - c(\theta)) f_0(x)$$

satisfies the standardized model with

$$c(\theta) = -(\theta + \ln(1 - \theta)),$$

and

$$\theta_i = 1 - \lambda_i/\lambda^*$$

for $i = 0, 1$ such that

$$c(\theta_0) = c(\theta_1).$$

A scatterplot or by looking at the data directly shows that there is a change around the observation 50. So we take the mean from observations 1 to 50 as the pre-change mean and the mean from observations 51 to 109 as the post-change mean, which gives

$$\lambda_0 = 1/129 \quad \text{and} \quad \lambda_1 = 1/335,$$

and $\lambda^* = 0.005$.

Now, after the data transformation by letting $x_i = 0.005y_i - 1$ for $i = 1, ...109$, we can fit the data into the standardized model with

$$\theta_0 = -0.550 \quad and \quad \theta_1 = 0.403.$$

Next, we can formalize the CUSUM process by calculating T_n's based on x_i's, which are reported as follows:

[1] 0.88812755 0.06794922 0.00000000 0.00000000 0.07393498
[6] 0.00000000 0.00000000 0.00000000 0.00000000 0.00000000
[11] 0.00000000 0.00000000 0.00000000 0.00000000 0.01399443
[16] 0.00000000 0.00000000 0.00000000 0.00000000 0.57343963
[21] 0.00000000 0.00000000 0.00000000 0.00000000 0.00000000
[26] 0.72329102 0.00000000 0.00000000 0.42858328 0.00000000

[31] 0.00000000 0.00000000 0.16384582 0.00000000 0.00000000
[36] 0.00000000 0.00000000 0.00000000 0.62838514 1.00202291
[41] 0.27175541 0.35568049 0.00000000 0.00000000 0.00000000
[46] 0.00000000 0.80321176 1.36166625 2.12991269 1.41962538
[51] 1.79326315 1.18287677 0.26779256 5.28682351 7.50363341
[56] 8.83632010 12.18700554 11.42676777 11.04115848 12.32389470
[61] 13.81142782 13.05618510 12.71053618 12.61963463 12.89337147
[66] 12.86740552 12.98629592 14.81349220 15.76156031 15.12120366
[71] 15.26007424 15.61373183 15.65270148 17.23514049 24.29215038
[76] 23.56188289 24.19026803 29.74376895 30.48204510 33.20335470
[81] 33.28727977 32.88668534 33.26032311 32.36022404 31.68989711
[86] 32.14345562 31.16343580 32.00660794 32.69493363 33.37326923
[91] 32.46817511 33.11154539 33.75991071 34.31836520 34.17251814
[96] 33.89679987 33.27142836 34.08962526 33.27444197 32.36934786
[101] 32.14857510 31.38334228 31.02770327 38.16962896 37.31448530
[106] 37.39841038 36.43337570 35.52328654 41.30156455

By inspecting the CUSUM process, we see that the estimator $\hat{\nu} = 46$ no matter what the threshold d is (at least 5 and at most 40). Again, we see that the CUSUM procedure is very reliable in terms of the change-point estimator.

2.4.3   IBM Stock Price (Variance Change)

Suppose the original independent observations $\{Y_i\}$ follow $N(0, \sigma_0^2)$ for $i \le \nu$ and $N(0, \sigma^2)$ for $i > \nu$.

For a reference value σ_1^2 for σ^2, we define

$$\lambda^* = \left(\frac{1}{\sigma_0^2} - \frac{1}{\sigma_1^2} \right) / \ln \left(\frac{\sigma_1^2}{\sigma_0^2} \right)$$

and make the following data transformation:

$$X_i = \frac{1}{\sqrt{2}}(\lambda^* Y_i^2 - 1).$$

Let $f_0(x)$ be the density function of $(\chi^2 - 1)/\sqrt{2}$, where χ^2 is the standard chi-square random variable with degree of freedom 1. Then from Example 1.3, we know

$$c(\theta) = -\frac{\theta}{\sqrt{2}} - \frac{1}{2}\ln(1 - \sqrt{2}\theta).$$

Define

$$\theta_0 = \frac{1}{\sqrt{2}}\left(1 - \frac{1/\sigma_0^2}{\lambda^*} \right) \quad \text{and} \quad \theta_1 = \frac{1}{\sqrt{2}}\left(1 - \frac{1/\sigma_1^2}{\lambda^*} \right),$$

and generally

$$\theta = \frac{1}{\sqrt{2}}\left(1 - \frac{1/\sigma^2}{\lambda^*} \right).$$

It can be verified that

$$c'(\theta) = \frac{1}{\sqrt{2}}\left(\frac{\lambda^*}{1/\sigma^2} - 1\right)$$

and $c(\theta_0) = c(\theta_1)$. Thus, the standardized observations $\{X_i\}$ fit in our model.

Remark: Generally, the original independent observations $\{Y_i\}$ follow $(1 + \epsilon_0)^2\chi_p^2$ for $1 \leq \nu$ and $(1 + \epsilon_1)^2\chi_p^2$ for $i > \nu$, where χ_p^2 is the standard chi-square random variable with degree of freedom p. Then, by using Example 1.3, we define

$$\lambda^* = \left(\frac{1}{(1 + \epsilon_0)^2} - \frac{1}{(1 + \epsilon_1)^2}\right)/\ln[(1 + \epsilon_1)^2/(1 + \epsilon_0)^2],$$

and make the following transformation:

$$X_i = \frac{1}{\sqrt{2p}}(\lambda^*Y_i - p).$$

For $f_0(x)$ defined as in Example 1.3, let

$$\theta = \sqrt{p/2}\left(1 - \frac{(1 + \epsilon)^2}{\lambda^*}\right),$$

and define θ_0 and θ_1 correspondingly. Then we can verify that $c(\theta_0) = c(\theta_1)$.

The following data set is taken as the IBM stock daily closing prices from May 17 of 1961 to Nov. 2 of 1962, [Box, Jenkins, and Reinsel(1994), pp.542)] for a total of 369 observations. The data are read in **rows**.

We use the geometric normal random walk model and find that it fits the data quite well with quite small autocorrelation. After taking the difference for the logarithm of the data, the plot of total 368 data shows an obvious increase in the variance roughly around the 225th observation.

The standard deviation for the first 225 observations is found to be 0.00978. So we divide all the data by 0.00978, and denote the modified data as $\{Y_i\}$'s, which gives

$$\sigma_0^2 = 1 \quad \text{and} \quad \sigma_1^2 = 7.239,$$

where σ_1^2 is the variance of the last 143 modified observations.

Now, we calculate λ^* as

$$\lambda^* = \left(\frac{1}{\sigma_0^2} - \frac{1}{\sigma_1^2}\right)/\ln\left(\frac{\sigma_1^2}{\sigma_0^2}\right) = (1 - 1/7.239)/\log(7.239) = 0.4354.$$

The transformed data are calculated as

$$X_i = \frac{1}{\sqrt{2}}(\lambda^*Y_i^2 - 1),$$

and the corresponding conjugate parameters are found to be

$$\theta_0 = \frac{1}{\sqrt{2}}\left(1 - \frac{1}{0.4354}\right) = -0.917$$

Table 2.4: IBM Stock Price

460	457	452	459	462	459	463	479
493	490	492	498	499	497	496	490
489	478	487	491	487	482	479	478
479	477	479	475	479	476	476	478
479	477	476	475	475	473	474	474
474	465	466	467	471	471	467	473
481	488	490	489	489	485	491	492
494	499	498	500	497	494	495	500
504	513	511	514	510	509	515	519
523	531	547	551	547	551	547	541
545	549	545	549	547	543	540	539
532	517	527	540	542	538	541	541
547	553	559	557	557	560	571	571
569	575	580	584	585	590	599	603
599	596	585	587	587	581	583	592
596	596	595	598	598	595	595	592
588	582	576	578	589	585	580	579
584	581	581	577	577	578	580	586
583	581	576	571	575	575	573	577
582	584	579	572	577	571	560	549
556	557	563	564	567	561	559	553
553	553	547	550	544	541	532	525
542	555	558	551	551	552	553	557
557	548	547	545	545	539	539	535
537	535	536	537	543	548	546	547
548	549	553	552	551	550	553	554
551	551	545	547	547	537	539	538
533	525	513	510	521	521	521	523
516	511	518	517	520	519	519	519
518	513	499	485	454	462	473	482
486	475	459	451	453	446	455	452
457	449	450	435	415	398	399	361
383	393	385	360	364	365	370	374
359	335	323	306	333	330	336	328
316	320	332	320	333	344	339	350
351	350	345	350	359	375	379	376
382	370	365	367	372	373	363	371
369	376	387	387	376	385	385	380
373	382	377	376	379	386	387	386
389	394	393	409	411	409	408	393
391	388	396	387	383	388	382	384
382	383	383	388	395	392	386	383
377	364	369	355	350	353	340	350
349	358	360	360	366	359	356	355
367	357	361	355	348	343	330	340
339	331	345	352	346	352	357	

and

$$\theta_1 = \frac{1}{\sqrt{2}} \left(1 - \frac{1}{7.239 * 0.4354} \right) = 0.483.$$

The CUSUM process formed by the standardized $\{X_i\}$'s are calculated as

[1] 0.00 0.00 0.05 0.00 0.00 0.00 3.01 4.97 4.39 3.73 3.50 2.80
[13] 2.15 1.45 1.22 0.53 1.49 1.90 1.41 0.92 0.55 0.00 0.00 0.00
[25] 0.00 0.00 0.00 0.00 0.00 0.00 0.00 0.00 0.00 0.00 0.00 0.00
[37] 0.00 0.00 0.00 0.00 0.48 0.00 0.00 0.00 0.00 0.00 0.00 0.20
[49] 0.16 0.00 0.00 0.00 0.00 0.00 0.00 0.00 0.00 0.00 0.00 0.00
[61] 0.00 0.00 0.00 0.00 0.30 0.00 0.00 0.00 0.00 0.00 0.00 0.00
[73] 0.00 0.00 0.04 2.16 1.63 1.09 0.78 0.24 0.00 0.00 0.00 0.00
[85] 0.00 0.00 0.00 0.00 1.93 2.40 3.60 2.94 2.41 1.80 1.10 0.78
[97] 0.46 0.12 0.00 0.00 0.00 0.51 0.00 0.00 0.00 0.00 0.00 0.00
[109] 0.00 0.03 0.00 0.00 0.00 0.41 0.00 0.00 0.00 0.00 0.05 0.00
[121] 0.00 0.00 0.00 0.00 0.00 0.00 0.00 0.00 0.00 0.00 0.00 0.44
[133] 0.00 0.00 0.00 0.00 0.00 0.00 0.00 0.00 0.00 0.00 0.00 0.00
[145] 0.00 0.00 0.00 0.00 0.00 0.00 0.00 0.00 0.00 0.00 0.00 0.00
[157] 0.00 0.51 1.07 0.88 0.18 0.00 0.00 0.00 0.00 0.00 0.00 0.00
[169] 0.00 0.00 0.00 0.00 0.00 0.20 0.06 2.62 3.72 3.11 2.91 2.21
[181] 1.51 0.81 0.27 0.00 0.15 0.00 0.00 0.00 0.00 0.00 0.00 0.00
[193] 0.00 0.00 0.00 0.00 0.00 0.00 0.00 0.00 0.00 0.00 0.00 0.00
[205] 0.00 0.00 0.00 0.00 0.00 0.00 0.00 0.00 0.00 0.39 0.00 0.00
[217] 0.00 0.03 1.04 0.45 1.21 0.50 0.00 0.00 0.00 0.00 0.00 0.00
[229] 0.00 0.00 0.00 0.00 0.00 0.00 1.76 3.66 16.99 17.28 18.34 18.80
[241] 18.29 19.27 22.34 22.63 21.99 22.06 22.64 22.07 21.76 22.05 21.36 24.35
[253] 30.78 35.70 35.01 66.55 77.11 78.54 79.19 92.99 92.68 92.00 91.89 91.55
[265] 96.24 110.94 114.52 123.22 145.53 145.08 145.42 146.58 150.35 150.15 153.80 157.46
[277] 161.12 164.77 169.17 171.86 171.84 174.42 173.74 173.06 173.02 172.98 187.59 187.24
[289] 186.74 186.84 189.41 189.30 188.69 188.57 187.89 189.56 190.38 189.77 190.20 192.17
[301] 191.46 193.43 194.52 193.82 193.66 194.06 195.19 195.04 194.35 193.85 194.22 193.54
[313] 192.85 192.34 192.15 191.47 195.89 195.26 194.63 193.94 197.75 197.12 196.61 197.24
[325] 198.24 197.88 197.71 197.78 197.17 196.55 195.86 195.15 194.99 195.31 194.79 194.85
[337] 194.34 194.43 197.69 197.58 201.69 201.63 201.16 204.98 206.98 206.30 207.68 207.07
[349] 206.36 206.54 207.30 206.55 205.87 208.72 210.47 210.16 210.36 210.93 210.89 214.99
[361] 217.15 216.47 217.60 222.42 223.01 223.26 223.50 223.43

We see that no matter what the threshold d will be (at least 5 and at most 223), the change-point estimator is consistently $\hat{\nu} = 234$.

The estimation for the post-change parameter is considered in Chapter 4.

3
Confidence Interval for Change-Point

3.1 A Lower Confidence Limit

In this chapter, we consider the construction of a lower confidence limit for the change-point after a change is detected. This may be interesting from a quality control point of view. When a signal is made, we want to inspect all the items after the change-point. Thus, a lower confidence limit is desirable to estimate the number of items to be inspected. Note that

$$\hat{\nu} = \max\{n < N : \ T_n = 0\},$$

that is, the last zero point of the CUSUM process T_n before the time to signal. The main difficulty in forming a confidence interval for the change-point in the sequential sampling case is the memory problem. In many practical problems, the change occurs quite far away from the starting time. Thus, it will not be feasible to memorize all the data up to the time of signal. Motivated from the renewal property of the CUSUM process T_n, we propose the following simple method to construct a lower confidence limit. We keep tracking the last s zero-points of T_n before the current time until the alarming time. More specifically, let L_k denote the kth last zero point of T_n counted backward starting from

$L_0 = \hat{\nu}$ for $k = 1, 2, ..., s$. Then L_s is a $(1 - \alpha)$-level lower confidence limit if

$$s = \inf\{k \geq 0 : \quad P^\nu(\nu < L_k | N > \nu) \leq \alpha\}, \tag{3.1}$$

that is, $[L_s, N)$ will be a lower confidence set for ν.

The advantage for this method is that one can update all related information recursively for any given value of s until the time to signal due to the renewal property of T_n. Also, we note that this is actually a conservative method. To obtain a more accurate lower confidence limit L^*, one can interpolate L_s and L_{s-1} with the plot of $\{P^\nu(\nu < L_k | N > \nu)\}$ for $k = 0, 1, 2, ..., s$.

To evaluate the noncoverage probability of $[L_s, N)$, that is,

$$p_s = P^\nu(\nu < L_s | N > \nu), \tag{3.2}$$

we first note that conditional on the value of T_ν given $N > \nu$, the zeropoints of T_n for $n > \nu$ will behave like a defective delayed renewal process and the total number of zero points after the first zero-point follows a geometric distribution with terminating probability $P_\theta(S_{N_0} \leq 0)$. Second, we note that the event $\{\nu < L_s\}$ is equivalent to the event that the number of zero points of T_n after ν is larger than or equal to $s + 1$. Thus, we have

$$p_s = P^\nu(S'_{N_{T_\nu}} \leq 0 | N > \nu) \left(P_\theta(S_{N_0} \leq 0)\right)^s, \tag{3.3}$$

where $\{S'_n\}$ is an independent copy of $\{S_n\}$ with $S'_0 = T_\nu$. In the next section, we shall give the asymptotic results for p_s and the average length of margin of error $E^\nu[\nu - L_s | L_s \leq \nu < N]$. Our discussion will be focused on the normal case. An extension to the exponential family case can be seen in Ding (2003) with slightly different forms.

3.2   Asymptotic Results

We assume that $\nu, d \to \infty$ conditional on $N > \nu$, that is, the quasistationary state. From Pollak and Siegmund (1986), we note that as $\nu, d \to \infty$,

$$P^\nu(T_\nu \in dx | N > \nu) \to P_\theta(M \in dx).$$

Thus,

$$P^\nu(S'_{N_{T_\nu}} \leq 0 | N > \nu) \to P_\theta(\tau_{-M} < \infty) = p_0.$$

Also, we can see that

$$P_\theta(S_{N_0} \leq 0) \to p = P_\theta(\tau_- < \infty).$$

Therefore, we have the following result using (3.3).

Theorem 3.1 *As $\nu, d \to \infty$, for any finite s,*

$$p_s \to p_0 p^s.$$

Thus, a conservative $(1 - \alpha)$-level lower confidence limit can be taken as L_{s^*}, where

$$s^* = [\ln(\alpha/p_0)/\ln p] + 1.$$

Next, we evaluate the average length of margin of error $E^\nu[\nu - L_s | L_s \leq \nu < N]$ as $\nu, d \to \infty$ conditional on that the change is covered.

Note that

$$E^\nu[\nu - L_s | L_s \leq \nu < N] = \frac{E^\nu[\nu - L_s; L_s \leq \nu | N > \nu]}{P^\nu(L_s \leq \nu | N > \nu)},$$

where $E[X; A] = E[X I_A]$, and

$$E^\nu[\nu - L_s; L_s \leq \nu | N > \nu]$$

$$= \sum_{k=s}^{1} E^\nu[\nu - L_s; L_k \leq \nu < L_{k-1} | N > \nu] + E^\nu[\nu - L_s; \hat\nu \leq \nu | N > \nu].$$

From an argument given in the proof of Theorem 3.1, we know that

$$E^\nu[\nu - \hat\nu; \hat\nu < \nu | N > \nu] \to E_{\theta_0}[\sigma_M P_\theta(\tau_{-M} = \infty)], \tag{3.4}$$

where

$$\sigma_M = \text{argmax}_{0 \leq k < \infty} S'_k$$

and

$$E^\nu[\hat\nu - L_s | \hat\nu \leq \nu \leq N] \to s E_{\theta_0}[S_{N_0} | S_{N_0} \leq 0] \to s E_{\theta_0}[\tau_-].$$

Thus,

$$E^\nu[\nu - L_s; \hat\nu < \nu | N > \nu] \to s E_{\theta_0}(\tau_-)(1 - p_0) + E_{\theta_0}[\sigma_M P_\theta(\tau_{-M} = \infty)].$$

Also, we see that, for $k = s, ..., 2, 1$,

$$E^\nu[\nu - L_s; L_k \leq \nu < L_{k-1} | N > \nu] = \sum_{j=s}^{k-1} \{ E^\nu[L_{j-1} - L_j | L_k \leq \nu < L_{k-1}; N > \nu]$$

$$+ E^\nu[\nu - L_k | L_k \leq \nu < L_{k-1}; N > \nu] \} P^\nu(L_k \leq \nu < L_{k-1} | N > \nu).$$

Now, for $j = s, ..., k - 1$,

$$E^\nu[L_{j-1} - L_j | L_k \leq \nu < L_{k-1}; N > \nu] \to E_{\theta_0}[S_{N_0} | S_{N_0} \leq 0] \to E_{\theta_0}[\tau_-]$$

and

$$P^\nu(L_k \leq \nu < L_{k-1} | N > \nu) \to p_0(1 - p)p^{k-1}.$$

Also, we can see that

$$E^\nu[\nu - L_k | L_k \leq \nu < L_{k-1}; N > \nu] \to E_{\theta_0 \theta}[\sigma_M | \tau_{-M} < \infty].$$

Combining the above results, we have

$$E^\nu[\nu - L_s; L_s \leq \nu | N > \nu]$$

$$\to \quad \sum_{k=s}^{1}(s-k)E_{\theta_0}(\tau_-)p_0(1-p)p^{k-1} + p_0(1-p^s)E_{\theta_0\theta}[\sigma_M | \tau_{-M} < \infty]$$
$$+ E_{\theta_0}[\sigma_M P_\theta(\tau_{-M} = \infty)] + s(1-p_0)E_{\theta_0}(\tau_-)$$
$$= \quad \left(s - p_0\frac{1-p^s}{1-p}\right)E_{\theta_0}(\tau_-) + E_{\theta_0}(\sigma_M) - p^s E_{\theta_0}[\sigma_M P_\theta(\tau_{-M} < \infty)].$$

In summary, we have the following result.

Theorem 3.2: *As $\nu, d \to \infty$,*

$$E^\nu[\nu - L_s | L_s \leq \nu < N] \to \frac{1}{1-p_0p^s}\{\{s - p_0\frac{1-p^s}{1-p}\}E_{\theta_0}(\tau_-) + E_{\theta_0}(\tau_{-M})$$

$$- p^s E_{\theta_0}[\sigma_M P_\theta(\tau_{-M} < \infty)]\}.$$

3.3   Second-Order Approximation

For numerical evaluation, we further assume that $\theta_0, \theta \to 0$ at the same order, which is lower than $O(1/d)$.

From Lemma 4 of Wu (1999), we know that

$$p = 1 - \sqrt{2}\theta e^{-\theta\rho} + O(\theta^3)$$

and

$$p_0 = \frac{-\theta_0}{\theta - \theta_0}e^{-\theta(\theta-\theta_0)} + O((\theta - \theta_0)^3),$$

where

$$\rho = \lim_{d\to\infty} E_0 R_d \approx 0.583.$$

Thus, we have by Theorem 3.1

$$p_s \approx \frac{-\theta_0}{\theta - \theta_0}e^{-\theta(\theta-\theta_0)}\left(1 - \sqrt{2}\theta e^{-\theta\rho}\right)^s. \tag{3.5}$$

At $\theta = -\theta_0$, Eq. (3.5) yields

$$p_s \approx \frac{1}{2}e^{-2\theta_0^2}\left(1 + \sqrt{2}\theta_0 e^{\theta_0\rho}\right)^s.$$

For the noncoverage probability $\alpha = p_s$, we can roughly find s as

$$s^* = -\ln\left(\frac{(\theta - \theta_0)\alpha}{-\theta_0}\right)\left(\frac{1}{\sqrt{2}\theta} - \frac{1-\sqrt{2}\rho}{2}\right) + o(\theta). \tag{3.6}$$

This approximation can give us a rough upper bound for s^* by taking $\theta - \theta_0$ as the smallest change magnitude we want to detect given the reference value $-\theta_0$. At $\theta = -\theta_0$, Eq. (3.6) gives

$$s^* = -\ln(2\alpha)\left(\frac{1}{-\theta_0\sqrt{2}} - 0.088\right) + o(\theta_0). \tag{3.7}$$

For example, at $\theta = -\theta_0 = 0.25$ and $\alpha = 0.05$, $s^* \approx 6.3$. In Table 3.1, a direct calculation based on (3.5) shows that at $s = 6$, $p_s \approx 0.05$, the same as the simulated value.

We now turn to the approximation for the average length of margin of error. A direct application of Lemma 10.27 of Siegmund (1985) gives

$$E_{\theta_0}(\tau_-) = \frac{1}{\theta_0}E_{\theta_0}S_{\tau_-}$$

$$= \frac{1}{-\sqrt{2}\theta_0}e^{-\theta_0\rho} + O(\theta_0).$$

Lemma 5 of Wu (1999) and Lemma 2.1 of Chapter 2 give, respectively,

$$E_{\theta_0}(\sigma_M) = \frac{1}{2\theta_0^2} + O(1)$$

and

$$E_{\theta_0}[\sigma_M P_\theta(\tau_{-M} < \infty)] = \frac{1}{2(\theta - \theta_0)^2} + O(1).$$

Substituting these approximations into Theorem 3.2, we obtain the approximated value as

$$E^\nu[\nu - L_s | L_s \le \nu < N]$$

$$\approx \frac{1}{1-\alpha}\left\{\frac{e^{-\theta_0\rho}}{-\sqrt{2}\theta_0}\left(s - \frac{-\theta_0}{\sqrt{2}\theta(\theta - \theta_0)}e^{\theta(\rho - (\theta - \theta - 0))}\right.\right.$$

$$\left.\times\left(1 - \frac{-(\theta - \theta_0)}{\theta_0}e^{(\theta - \theta_0)\theta}\alpha\right) + \frac{1}{2\theta_0^2} - \frac{\alpha}{-2\theta_0(\theta - \theta_0)}e^{(\theta - \theta_0)\theta}\right\},$$

where $\alpha = p_0 p^s$ denotes the non-coverage probability.

At $\theta = -\theta_0$,

$$E^\nu[\nu - L_s | L_s \le \nu < N]$$

$$\approx \frac{1}{1-\alpha}\left[\frac{e^{-\theta_0\rho}}{-\sqrt{2}\theta_0}\left(s - \frac{e^{-2\theta_0(\rho + 2\theta_0)/2}}{-2\sqrt{2}\theta_0}\left(1 - 2\alpha e^{-\theta_0}\right)\right) + \frac{1}{2\theta_0^2} - \frac{\alpha e^{-\theta_0}}{(2\theta_0)^2}\right]. \tag{3.8}$$

To check the accuracy of these approximations, we conducted a simulation study with 5000 replications. For average length

$$ARL_0 = E_{\theta_0}(N) = 1000$$

Table 3.1: Noncoverage Probabilities of Lower Confidence Limit

$-\theta_0$	θ	$s=0$	1	2	3	4	5	6	7	8
0.2	0.2	0.462	0.345	0.258	0.193	0.144	0.108	0.081	0.061	0.0
$d=$		0.460	0.341	0.251	0.184	0.139	0.105	0.078	0.057	0.0
9.96	0.45	0.397	0.276	0.192	0.133	0.092	0.064	0.045	0.031	0.0
		0.403	0.281	0.193	0.130	0.092	0.061	0.044	0.032	0.0
	0.25	0.344	0.222	0.143	0.092	0.059	0.038	0.025	0.016	0.0
		0.353	0.228	0.150	0.097	0.061	0.041	0.026	0.017	0.0
0.25	0.25	0.441	0.306	0.213	0.148	0.103	0.071	0.050	0.034	0.0
$d=$		0.452	0.312	0.215	0.148	0.105	0.072	0.050	0.035	0.0
8.59	0.55	0.385	0.248	0.160	0.103	0.066	0.043	0.027	0.018	
		0.401	0.254	0.164	0.104	0.065	0.044	0.028	0.018	
	0.35	0.338	0.201	0.120	0.072	0.043	0.025	0.015		
		0.369	0.213	0.124	0.070	0.039	0.020	0.013		
0.3	0.3	0.418	0.269	0.173	0.112	0.072	0.046	0.039	0.019	
$d=$		0.435	0.280	0.182	0.117	0.075	0.048	0.030	0.020	
7.56	0.65	0.368	0.219	0.131	0.078	0.047	0.028	0.017		
		0.387	0.229	0.140	0.084	0.049	0.031	0.018		
	0.4	0.324	0.179	0.099	0.055	0.030	0.017			
		0.347	0.197	0.108	0.058	0.030	0.017			

and $-\theta_0 = 0.2, 0.25, 0.3$, the design for the control limit d is obtained by using the approximation given in Chapter 1:

$$ARL_0 \approx \frac{2}{(2\theta_0)^2} \left(e^{-2\theta_0(d+2\rho)} - 1 + 2\theta_0(d + 2\rho) \right),$$

which gives corresponding values of d as 9.96, 8.59, and 7.56, respectively. The change-point is taken as $\nu = 100$ to ensure that the detecting probability is close to 1. Table 3.1 gives the comparison between the approximated and simulated noncoverage probabilities for several typical values of s. In each cell, the top number is the approximated value while the bottom number is the simulated value.

In Table 3.2, we also list the comparisons between the simulated average margin of error and approximated values given in (3.8). Again, the top number is the approximated value and the bottom number is the simulated value. We see that the approximations are slightly unsatisfactory, mainly due to the fact that we are unable to obtain the third-order expansions for $E[\sigma_M P[\tau_{-M} < \infty]]$.

3.4   Estimated Lower Limit

In most practical situations, the post-change mean is rarely known. Thus, we need to estimate θ after detection in order to find the estimated s^*.

Table 3.2: Average Margin of Error of Lower Confidence Limit

$\delta = \mu$	$s = 0$	1	2	3	4	5	6	7	8
0.4	17.43	19.26	20.93	22.56	24.21	25.88	27.59	29.34	31.12
	12.73	14.45	16.34	18.72	21.72	24.56	27.16	30.19	33.49
0.5	10.73	12.30	13.70	15.08	16.48	17.91	19.37		
	8.81	10.23	12.06	14.23	16.50	19.22	21.76		
0.6	7.16	8.53	9.76	10.97	12.20	13.47	14.76	16.08	
	6.03	7.22	8.98	11.08	13.25	15.46	17.94	20.39	

Note that $L_0 = \hat{\nu}$ is the maximum likelihood estimation of ν at $\theta = -\theta_0$. We naturally estimate θ by the sample mean after $\hat{\nu}$, that is,

$$\hat{\theta} = \frac{T_N}{N - \hat{\nu}}.$$

Under the same conditions as in Sections 4.2 and 4.3 of Chapter 4, the bias of $\hat{\theta}$ is approximately equal to

$$E^{\nu}[\hat{\theta} - \theta | N > \nu] \approx \frac{1}{d}\left(2 - \frac{(\theta/\theta_0)^3}{2(\theta/\theta_0 - 1)}\right),$$

which is valid for θ/θ_0 within a range.

Therefore, we can use the corrected estimator for θ, for example as in White-head (1986), by solving

$$\tilde{\theta} = \hat{\theta} - \frac{1}{d}\left(2 - \frac{(\hat{\theta}/\theta_0)^3}{2(\hat{\theta}/\theta_0 - 1)}\right),$$

under the restriction for the value of $\hat{\theta}/\theta_0$. By substituting $\tilde{\theta}$ into (3.5) for p_s, we can obtain the estimated value of s such that the noncoverage probability is approximately equal to the required significance level α. The investigation of this estimated lower confidence limit will involve the study on the performance of $\tilde{\theta}$. A detailed analysis is given in the next chapter.

3.5 Confidence Set

For detecting jump points of a stable process, it would also be interesting to provide a confidence set for the change point. In contrast to the finite sample size case, the data available from the change-point to the signal time are very limited.

Here we assume that the data after the last zero point of T_n until the current time are restored fully. That means, at the signal time N, the data $\{T_n\}$ for $\hat{\nu} \le n \le N$ are available. Here, we follow an approach used by Siegmund (1988) in the fixed sample size case.

Define a set

$$V_c = \{k \geq \hat{\nu} : \quad T_k \leq c\},$$

for $0 < c < d$, which does not necessarily give an interval. Then $V_c \cup [L_s, \hat{\nu})$ will be a $(1-\alpha)$-level confidence set if

$$P^\nu(\nu \leq L_s | N > \nu) = \frac{\alpha}{2} = P^\nu(T_v > c; \hat{\nu} \geq \nu | N > \nu).$$

From the analysis in Section 3.3, it is easy to see that as $\nu, d \to \infty$,

$$P^\nu(T_v > c; \hat{\nu} \geq \nu | N > \nu) \to P_{\theta_0 \theta}[M > c; \tau_{-M} = \infty].$$

For the design purpose, we give a rough second-order approximation as in Section 3.4. By using the Wald's likelihood ratio identity and approximating the average overshoot by 0.583, we have

$$
\begin{aligned}
P_{\theta_0 \theta}[M > c; \tau_{-M} = \infty] &= E_{\theta_0}[P_\theta(\tau_{-M} = \infty); M > c] \\
&= E_{\theta_0}\left[\left(1 - E_\theta e^{-2\theta S_{\tau_M}}\right); M > c\right] \\
&= E_{\theta_0}\left[\left(1 - e^{-2\theta(M+\rho)}\right); M > c\right] \\
&= -\int_c^\infty \left(1 - e^{-2\theta(x+\rho)}\right) dP_{\theta_0}(M > x) \\
&= \int_c^\infty \left(1 - e^{-2\theta(x+\rho)}\right) de^{2\theta_0(x+\rho)} \\
&= e^{2\theta_0(c+\rho)} - \frac{-\theta_0}{\theta - \theta_0} e^{-2\mu(c+\rho)}.
\end{aligned}
$$

The average length of the confidence set can be studied similarly to Sections 3.3 and 3.4. Finally, we should point out that an upper confidence limit can be defined as

$$U_c = \inf\{k \geq \hat{\nu} : \quad T_k \geq c\}$$

if a confidence interval is desired.

4

Inference for Post-Change Mean

In this chapter, we consider the inference problem for the post-change mean in the normal case. The technique can be extended to the more general exponential family case. The materials are basically taken from Wu (2004).

4.1 Inference for θ When $\hat{\nu} = \nu$

4.1.1 Estimation

For $X_1, ..., X_n, ...$ i.i.d. following $N(\theta, 1)$, we condition on $S_{N_x} > d$, that means the random walk S_n stays positive before it crosses the boundary d. Based on the samples S_n for $n \leq N_x$, the maximum likelihood estimator of θ is given by

$$\hat{\theta}_x = \frac{S_{N_x} - x}{N_x}.$$

It can be seen that when $x = 0$, $\hat{\theta}_0$ is indeed equivalent to the post-change mean estimator when $\hat{\nu} = \nu$. Our discussion in this section will be focused on $\hat{\theta}_0$, which provides the basic techniques and also has its own independent interest. The results can be generalized to $\hat{\theta}_x$ for any $0 \leq x < d$ and will be used in the next section.

4.1.2   Bias of $\hat{\theta}_0$

We first consider the bias of $\hat{\theta}_0$, as in Siegmund (1978) and Whitehead (1986) for the estimates after sequential testing.

The following two lemmas are essential to prove the main result.

Lemma 4.1: *As $d \to \infty$, for given $\theta > 0$,*

$$P_\theta(S_{N_0} > d) = P_\theta(\tau_- = \infty) + O(e^{-2\theta d}).$$

Proof: By using the total probability law, we can write

$$P_\theta(S_{N_0} > d) = P_\theta(\tau_- = \infty) + P_\theta(\tau_- < \infty; S_{N_0} > d).$$

Now, by conditioning on the value of S_{N_0} for $S_{N_0} > d$ and using the Wald's likelihood ratio identity [e.g. Siegmund (1985, Prop. 2.24)] by changing θ to $-\theta$, we have

$$P_\theta(\tau_- < \infty; S_{N_0} > d) = E_\theta[P_\theta(\tau_- < \infty | S_{N_0}); S_{N_0} > d]$$

$$= E_\theta\left[e^{-2\theta S_{N_0}} E_{-\theta} e^{2\theta R_{-\infty}}; S_{N_0} > d\right](1 + o(1))$$

$$= e^{-2\theta d} E_{-\theta} e^{2\theta R_{-\infty}} E_\theta\left[e^{-2\theta(S_{N_0} - d)}; S_{N_0} > d\right](1 + o(1))$$

$$= O(e^{-2\theta d}),$$

where in the second equation from the last, the limit overshoot is $R_{-\infty}$ by considering that the random walk goes below the boundary $-S_{N_0}$.

The lemma shows that the convergence is exponentially fast as $d \to \infty$. The next lemma gives the first-order approximation for the first two moments of N_0 given $S_{N_0} > d$. Since the proof is of independent interest, we give it in Section 4.5.1.

Lemma 4.2: *Conditional on $S_{N_0} > d$, as $d \to \infty$, for any $\theta > 0$,*

$$N_0 = \frac{d}{\theta}(1 + o_p(1)),$$

$$E_\theta[N_0 | S_{N_0} > d] = \frac{d}{\theta} + O(1),$$

$$\mathrm{Var}_\theta(N_0 | S_{N_0} > d) = \frac{d}{\theta^3} + O(1).$$

The following theorem gives the first-order bias of $\hat{\theta}_0$.

Theorem 4.1: *As $d \to \infty$, for any given $\theta > 0$,*

$$E_\theta[\hat{\theta}_0 - \theta | S_{N_0} > d] = \frac{a_0(\theta)}{d}(1 + o(1)),$$

where

$$a_0(\theta) = 1 + \theta \frac{\frac{\partial}{\partial \theta} P_\theta(\tau_- = \infty)}{P_\theta(\tau_- = \infty)}.$$

Proof: The proof generalizes the techniques used in Woodroofe (1990) and Coad and Woodroofe (1998) to the conditional expectation case. By using the basic property of normal density, we have

$$
\begin{aligned}
E_\theta[\hat\theta_0 - \theta; S_{N_0} > d] &= E_\theta\left[\frac{S_{N_0} - \theta N_0}{N_0}; S_{N_0} > d\right] \\
&= \frac{\partial}{\partial \theta} E_\theta\left[\frac{1}{N_0}; S_{N_0} > d\right].
\end{aligned}
$$

We first write

$$E_\theta\left[\frac{1}{N_0}; S_{N_0} > d\right] = \frac{\theta}{d} P_\theta(S_{N_0} > d) + E_\theta\left[\left(\frac{1}{N_0} - \frac{\theta}{d}\right); S_{N_0} > d\right].$$

From Lemma 4.1 and the proof of Lemma 4.2, we note that

$$\frac{\partial}{\partial \theta}\left(\frac{\theta}{d} P_\theta(S_{N_0} > d)\right) = \frac{1}{d}\frac{\partial}{\partial \theta}(\theta P_\theta(\tau_- = \infty))(1 + o(1)).$$

Next, we have

$$
\begin{aligned}
\frac{\partial}{\partial \theta} E_\theta\left[\left(\frac{1}{N_0} - \frac{\theta}{d}\right); S_{N_0} > d\right] &= \frac{\partial}{\partial \theta} E_0\left[\left(\frac{1}{N_0} - \frac{\theta}{d}\right) e^{\theta S_{N_0} - \theta^2 N_0/2}; S_{N_0} > d\right] \\
&= E_\theta\left[\left(\frac{1}{N_0} - \frac{\theta}{d}\right)(S_{N_0} - \theta N_0) - \frac{1}{d}; S_{N_0} > d\right] \\
&= E_\theta\left[\frac{(N_0 - d/\theta)^2}{d^2/\theta^3} - \frac{1}{d}; S_{N_0} > d\right](1 + o(1)) \\
&= \frac{1}{d} E_\theta\left[\frac{(N_0 - d/\theta)^2}{d/\theta^3} - 1; S_{N_0} > d\right](1 + o(1)) \\
&= o\left(\frac{1}{d}\right),
\end{aligned}
$$

from Lemma 4.2. The result is obtained by summarizing the above derivations. Note that [see for example, Lemma 8.23 of Siegmund (1985)]

$$P_\theta(\tau_- = \infty) = \frac{1}{E_\theta \tau_+} = \frac{\theta}{E_\theta S_{\tau_+}}.$$

Thus, we can write

$$
\begin{aligned}
E_\theta\left[\frac{S_{N_0}}{N_0} - \theta \mid S_{N_0} > d\right] &= \frac{1}{d}\left(1 + E_\theta S_{\tau_+} \frac{\partial}{\partial \theta}(\theta/E_\theta(S_{\tau_+}))\right)(1 + o(1)) \\
&= \frac{1}{d}\left(2 - \theta \frac{\partial}{\partial \theta} \ln E_\theta(S_{\tau_+})\right)(1 + o(1)).
\end{aligned}
$$

For a numerical evaluation, we consider the local approximation as $\theta \to 0$. Detailed techniques are referred to Chapter 2 where the bias of change-point estimation $\hat{\nu}$ is considered. To save space, we only borrow some results.

Locally, as $\theta \to 0$, Lemmas 4.1 and 4.2 are still valid if

$$\theta d^{1-\epsilon} \to \infty$$

for some $\epsilon > 0$. From Lemma 10.27 of Siegmund (1985),

$$E_\theta S_{\tau_+} = E_0 S_{\tau_+} e^{\theta \rho}(1 + o(\theta^2)),$$

where

$$\rho = E_0 R_\infty = E_0 S_{\tau_+}^2/(2 E_0 S_{\tau_+}) \approx 0.583.$$

Thus, we have the following local approximation.

Corollary 4.1: *As $\theta \to 0$ and $\theta d^{1-\epsilon} \to \infty$ for some $\epsilon > 0$, the first-order bias is approximately*

$$E_\theta[\hat{\theta}_0 - \theta | S_{N_0} > d] = \frac{1}{d}(2 - \theta \rho + o(\theta^2))(1 + o(1)). \tag{4.1}$$

The corresponding result for general $S_0 = x$ will be used extensively for estimated $\hat{\nu}$, which is given in the following corollary.

Corollary 4.2: *As $d \to \infty$, for any fixed $x \geq 0$,*

$$E_\theta[\hat{\theta}_x - \theta | S_{N_x} > d] = \frac{1}{d}\left(1 + \theta \frac{\frac{\partial}{\partial \theta} P_\theta(\tau_{-x} = \infty)}{P_\theta(\tau_{-x} = \infty)}\right)(1 + o(1)).$$

4.1.3   A Corrected Normal Pivot

We first give the following normality result.

Lemma 4.3: *As $d \to \infty$, for any $\theta > 0$, conditional on $S_{N_0} > d$,*

$$Z_{N_0} = \sqrt{N_0}\left(\frac{S_{N_0}}{N_0} - \theta\right)$$

weakly converges to a standard normal random variable.

Proof: Similar to Lemma 4.1, using the total probability law, we first write

$$P_\theta(Z_{N_0} < z; S_{N_0} > d) = P_\theta(Z_{\tau_d} < z) - P_\theta(Z_{\tau_d} < z; S_{N_0} \leq 0).$$

From Anscombe's theorem [Theorem 2.40 of Siegmund(1985)], we know as $d \to \infty$,

$$P_\theta(Z_{\tau_d} < z) \to \Phi(z),$$

where $\Phi(z)$ is the standard normal distribution function. On the other hand, by conditioning the value of S_{N_0}, we can write

$$P_\theta(Z_{\tau_d} < z; S_{N_0} \leq 0) = E_\theta[P_\theta(Z_{\tau_d} < z|S_{N_0}); S_{N_0} \leq 0].$$

Given $S_{N_0} \leq 0$, conditioning on $N_0 = n_0$ and $S_{N_0} = s$, τ_d is equivalent to $n_0 + \tau_{d-s}$ in distribution and S_{τ_d} is equivalent to $S_{\tau_{d-s}} + s$. Thus, as $d \to \infty$, we have

$$
\begin{aligned}
P_\theta(Z_{\tau_d} < z|N_0 = n_0; S_{N_0} = s) &= P_\theta\left((n_0 + \tau_{d-s})^{1/2}\left(\frac{S_{\tau_{d-s}} + s}{n_0 + \tau_{d-s}} - \theta\right) < z\right) \\
&= P_\theta(Z_{\tau_{d-s}}(1 + o_p(1)) < z) \to \Phi(z),
\end{aligned}
$$

uniformly for any finite number of n_0's and s in a finite closed negative interval. The lemma is proved after some standard treatment for the random N_0 and S_{N_0} given $S_{N_0} \leq 0$.

In the following, we extend Woodroofe (1992)'s method to find a corrected asymptotic normal pivot based on Z_{N_0} by obtaining the first-order bias of Z_{N_0} and the variance after bias correction. A similar idea has been discussed extensively in Coad and Woodroofe (1996), Woodroofe and Coad (1997), and Whitehead, Todd, and Hall (2000). Our main contribution here is the extension to the conditional expectation case. The following theorem gives the bias for Z_{N_0}.

Theorem 4.2: *As $d \to \infty$, for $\theta > 0$,*

$$E_\theta[Z_{N_0}|S_{N_0} > d] = \frac{b_0(\theta)}{\sqrt{d}}(1 + o(1)),$$

where

$$b_0(\theta) = \frac{1}{2\sqrt{\theta}} + \sqrt{\theta}\frac{\frac{\partial}{\partial\theta}P_\theta(\tau_- = \infty)}{P_\theta(\tau_- = \infty)}.$$

Proof: By generalizing Theorem 2 of Woodroofe (1990) and using the similar technique given in Theorem 1, we have

$$
\begin{aligned}
E_\theta[Z_{N_0}; S_{N_0} > d] &= \frac{\partial}{\partial\theta}E_\theta\left[\frac{1}{\sqrt{N_0}}; S_{N_0} > d\right] \\
&= \frac{1}{\sqrt{d}}\frac{\partial}{\partial\theta}(\theta^{1/2}P_\theta(S_{N_0} > d))(1 + o(1)) \\
&= \frac{1}{\sqrt{d}}\frac{\partial}{\partial\theta}(\theta^{1/2}P_\theta(\tau_- = \infty))(1 + o(1)).
\end{aligned}
$$

By using the same argument for proving Corollary 4.1, we note that as $\theta \to 0$,

$$\ln P_\theta(\tau_- = \infty) = \ln\left(\theta e^{-\theta\rho}/E_0 S_{\tau_+}\right)(1 + o(1)).$$

Thus, we have the following local approximation.

Corollary 4.3: *Locally as $\theta \dot{\rightarrow} 0$ and $\theta d^{1-\epsilon} \rightarrow \infty$ for some $\epsilon > 0$,*

$$E_\theta[Z_{N_0}|S_{N_0} > d] = \frac{1}{\sqrt{d\theta}}\left(\frac{3}{2} - \theta\rho + o(\theta^2)\right)(1 + o(1)). \tag{4.2}$$

To complete the construction of the corrected pivot, we need to find the variance of $Z_{N_0} - b_0(\hat{\theta}_0)/d^{1/2}$, which is equivalent to the second moment up to the order $O(1/d)$.

Theorem 4.3: *As $d \rightarrow \infty$,*

$$E_\theta[(Z_{N_0} - b_0(\hat{\theta}_0)/d^{1/2})^2|S_{N_0} > d] = 1 + \frac{c_0(\theta)}{d}(1 + o(1)),$$

where

$$c_0(\theta) = \frac{1}{4\theta} - \theta\frac{\partial^2}{\partial^2\theta}\ln P_\theta(\tau_- = \infty).$$

Proof: First, we write

$$E_\theta[(Z_{N_0} - b_0(\hat{\theta}_0)/d^{1/2})^2|S_{N_0} > d] = E_\theta[Z_{N_0}^2|S_{N_0} > d]$$

$$- \frac{2}{d^{1/2}}E_\theta[Z_{N_0}b_0(\hat{\theta}_0)|S_{N_0} > d] + \frac{1}{d}E_\theta[b_0^2(\hat{\theta}_0)|S_{N_0} > d]$$

$$= E_\theta[Z_{N_0}^2|S_{N_0} > d] - \frac{b_0^2(\theta)}{d} - \frac{2}{d^{1/2}}E_\theta[Z_{N_0}(b_0(\hat{\theta}_0) - b_0(\theta))|S_{N_0} > d](1 + o(1)).$$

For the first term on the right-hand side, we generalize Theorem 3.2 of Woodroofe (1990) by using the basic property of normal density and the technique in Theorem 4.1 and have

$$\begin{aligned}
E_\theta[Z_{N_0}^2; S_{N_0} > d] &= \frac{\partial^2}{\partial^2\theta}E_\theta\left[\frac{1}{N_0}; S_{N_0} > d\right] + P_\theta(S_{N_0} > d) \\
&= \frac{\partial^2}{\partial^2\theta}\left(\frac{\theta}{d}P_\theta(S_{N_0} > d)\right)(1 + o(1)) + P_\theta(S_{N_0} > d) \\
&= P_\theta(\tau_- = \infty) + \frac{1}{d}\frac{\partial^2}{\partial^2\theta}(\theta P_\theta(\tau_- = \infty))(1 + o(1)).
\end{aligned}$$

For the third term, the first-order Taylor expansion of $b_0(\hat{\theta}_0)$ at θ gives

$$\begin{aligned}
E_\theta[Z_{N_0}(b_0(\hat{\theta}_0) - b_0(\theta))|S_{N_0} > d] &= E_\theta[\sqrt{N_0}(\hat{\theta}_0 - \theta)^2 b_0'(\theta)|S_{N_0} > d](1 + o(1)) \\
&= \frac{b_0'(\theta)\theta^{1/2}}{d^{1/2}}E_\theta[Z_{N_0}^2|S_{N_0} > d](1 + o(1)) \\
&= \frac{b_0'(\theta)\theta^{1/2}}{d^{1/2}}(1 + o(1)),
\end{aligned}$$

which gives the expected result after some algebraic simplifications. Locally, as $\theta \rightarrow 0$, using the similar argument in Corollary 4.1, we see that

$$\frac{\partial^2}{\partial\theta^2}\ln P_\theta(\tau_- = \infty) = -\frac{1}{\theta^2}(1 + o(1)).$$

Combining with Corollary 4.2, we have

Corollary 4.4: *Locally as $\theta \to 0$, and $\theta d^{1-\epsilon} \to \infty$ for some $\epsilon > 0$,*

$$E_\theta[(Z_{N_0} - b_0(\hat\theta_0)/\sqrt{d})^2|S_{N_0} > d] = 1 + \frac{5}{4d}\left(\frac{1}{\theta} + o(1)\right). \qquad (4.3)$$

A corrected normal pivot can be formed as

$$Z^*_{N_0} = \frac{Z_{N_0} - b_0(\hat\theta_0)/d^{1/2}}{(1 + c_0(\hat\theta_0)/d)^{1/2}},$$

which gives a normal approximation accurate up to the order $o(1/d)$.

A $(1 - \alpha)$-level corrected confidence interval for θ can be constructed as

$$\hat\theta_0 - \frac{b_0(\hat\theta_0)}{d^{1/2}N_0^{1/2}} \pm \frac{(1 + c_0(\hat\theta_0)/d)^{1/2}}{N_0^{1/2}}z_{\alpha/2},$$

where z_α is the $(1 - \alpha)$th quantile of standard normal density.

Consequently, a bias-corrected estimation for θ can be obtained as

$$\theta^*_0 = \hat\theta_0 - \frac{b_0(\hat\theta_0)}{d^{1/2}N_0^{1/2}},$$

which is the approximate median and has a local correction $3/(2d)$ as $\theta \to 0$ as given in Corollary 4.2.

An alternative bias-corrected estimator $\tilde\theta_0$ can be obtained by solving

$$\tilde\theta_0 = \hat\theta_0 - a_0(\tilde\theta_0)/d,$$

as proposed in Whitehead(1986), which is approximately equal to

$$\tilde\theta_0 = \hat\theta_0 - \frac{a_0(\hat\theta_0)/d}{1 + a_0'(\hat\theta_0)/d},$$

with a local bias correction of $2/d$ as $\theta \to 0$ given in Corollary 4.1.

4.2   Post-Change Mean

In this section, we adapt the results and techniques of Section 4.1 to the estimate $\hat\theta$ when ν is estimated by $\hat\nu$.

Let $\nu \to \infty$ conditioning on $N > \nu$. That means, we consider the quasistationary state by assuming a change has been detected. As $d \to \infty$, from Pollak and Siegmund (1986),

$$\lim_{\nu\to\infty} P_{\theta_0}(T_\nu < y|N > \nu) \quad \to \quad \lim_{\nu\to\infty} P_{\theta_0}(T_\nu < y)$$
$$= \quad P_{\theta_0}(M < y),$$

where

$$M = \sup_{k \geq 0} S'_k,$$

and $\{S'_n\}$ is an independent copy of $\{S_n\}$ with drift θ_0 and $S'_0 = 0$. That means, the quasistationary state distribution of the CUSUM process is asymptotically equivalent to its stationary state distribution.

4.2.1   Bias of $\hat{\theta}$

We first study the bias of $\hat{\theta} = T_N/(N - \hat{\nu})$. By considering whether $\{\hat{\nu} \leq \nu\}$ or $\{\hat{\nu} > \nu\}$, we can write

$$E^\nu[\hat{\theta} - \theta | N > \nu] = E^\nu[\hat{\theta} - \theta; \hat{\nu} > \nu | N > \nu] + E^\nu[\hat{\theta} - \theta; \hat{\nu} \leq \nu | N > \nu].$$

As $\nu \to \infty$ and d approaches ∞, the event $\{\hat{\nu} > \nu\}$ is asymptotically equivalent to the event $\{\tau_{-M} = \infty\}$. Given $\hat{\nu} > \nu$, we note that $\hat{\theta}$ is equivalent to S_{N_0}/N_0 given $S_{N_0} > d$ in distribution as $\nu \to \infty$.

On the other hand, given $\hat{\nu} \leq \nu$, $\hat{\theta}$ is asymptotically equivalent to $S_{N_M}/(N_M + \sigma_M)$, where σ_M is the maximum point of $\{S'_n\}$, which is the asymptotic distance $\nu - \hat{\nu}$.

Thus, as $\nu, d \to \infty$,

$$
\begin{aligned}
E^\nu[\hat{\theta} - \theta | N > \nu] &= E_\theta[\hat{\theta}_0 - \theta | S_{N_0} > d] P_{\theta_0\theta}(\tau_{-M} < \infty)(1 + o(1)) \\
&\quad + E_{\theta_0\theta}\left[\frac{S_{N_M}}{N_M + \sigma_M} - \theta; S_{N_M} > d\right](1 + o(1)). \quad (4.4)
\end{aligned}
$$

The first expectation of (4.4) is already studied in Theorem 4.1 of Section 4.2. For the second expectation of (4.4), we have the following lemma:

Lemma 4.4: *As $d \to \infty$,*

$$E_{\theta_0\theta}\left[\frac{S_{N_M}}{N_M + \sigma_M} - \theta; S_{N_M} > d\right]$$

$$= \frac{1}{d}\left[\frac{\partial}{\partial\theta}E_\theta(\theta P_\theta(\tau_{-M} = \infty)) + \theta E_{\theta_0}[(M - \theta\sigma_M)P_\theta(\tau_{-M} = \infty)]\right](1 + o(1)).$$

Proof. By noting

$$\hat{\theta}_M = (S_{N_M} - M)/N_M,$$

we first write

$$E_{\theta_0\theta}\left[\frac{S_{N_M}}{N_M + \sigma_M} - \theta; S_{N_M} > d\right]$$

$$
\begin{aligned}
&= E_{\theta_0\theta}\left[\frac{S_{N_M} - M}{N_M + \sigma_M} - \theta; S_{N_M} > d\right] + E_{\theta_0\theta}\left[\frac{M}{N_M + \sigma_M}; S_{N_M} > d\right] \\
&= E_{\theta_0\theta}\left[\hat{\theta}_M\left(1 - \frac{\sigma_M}{N_M}\right) - \theta; S_{N_M} > d\right] \\
&\quad + E_{\theta_0\theta}\left[\frac{M}{N_M}\left(1 - \frac{\sigma_M}{N_M}\right); S_{N_M} > d\right](1 + o(1))
\end{aligned}
$$

$$= E_{\theta_0\theta}[\hat{\theta}_M - \theta; S_{N_M} > d] + E_{\theta_0\theta}\left[\frac{1}{N_M}(M - \hat{\theta}_M\sigma_M); S_{N_M} > d\right](1 + o(1)). \quad (4.5)$$

By using the result of Corollary 4.2 by treating M as x, the first term of (4.5) can be obtained as

$$E_{\theta_0\theta}\left[\frac{S_{N_M} - M}{N_M} - \theta; S_{N_M} > d\right] = \frac{1}{d}E_{\theta_0}\left(\frac{\partial}{\partial\theta}(\theta P_\theta(\tau_{-M} = \infty))\right)(1 + o(1)).$$

For the second term of (4.5), given M, using Lemmas 4.1 and 4.2, we get the following facts:

$$\hat{\theta}_M = \theta(1 + o_p(1)) \quad \text{and} \quad N_M = \frac{d}{\theta}(1 + o_p(1)),$$

which finishes the proof. Combining Theorem 4.1 and Lemma 4.4, we have

Theorem 4.4: *As $\nu \to \infty$ and d approaches ∞, for $\theta > 0$,*

$$E^\nu[\hat{\theta} - \theta|N > \nu] = \frac{a(\theta)}{d}(1 + o(1)),$$

where

$$\begin{aligned}
a(\theta) &= 1 + \theta\frac{\partial}{\partial\theta}P_{\theta_0\theta}(\tau_{-M} = \infty) + \theta\frac{P_{\theta_0\theta}(\tau_{-M} < \infty)}{P_\theta(\tau_- = \infty)}\frac{\partial}{\partial\theta}P_\theta(\tau_- = \infty) \\
&\quad + \theta E_{\theta_0}[(M - \theta\sigma_M)P_\theta(\tau_{-M} = \infty)].
\end{aligned}$$

Locally, as $\theta_0 \to 0$ at the same order, the renewal theorem shows that, in distribution,

$$-2\theta_0 M \to Y,$$

which follows the standard negative exponential distribution. Indeed, by using Wald's likelihood ratio identity, we have

$$\begin{aligned}
P_{\theta_0}(M < -\frac{x}{2\theta_0}) &= P_{\theta_0}(\tau_{-x/2\theta_0} = \infty) \\
&= 1 - P_{\theta_0}(\tau_{-x/2\theta_0} < \infty) \\
&= 1 - E_{-\theta_0}e^{2\theta_0(-x/(2\theta_0)+R_\infty)}(1 + o(1)) \\
&= 1 - e^{-x}(1 + o(1)).
\end{aligned}$$

Thus, by using Wald's likelihood ratio identity and the symmetric property of normal density, we have

$$\begin{aligned}
P_{\theta_0\theta}(\tau_{-M} < \infty) &= E_{\theta_0}[P_\theta(\tau_{-M} < \infty)] \\
&= E_{\theta_0}\left[E_\theta e^{-2\theta S_{\tau_M}}\right] \\
&= E_{\theta_0}e^{-2\theta M}(1 + o(1)) \\
&= Ee^{\frac{\theta}{\theta_0}Y}(1 + o(1)) \\
&= \frac{1}{1 - \theta/\theta_0}(1 + o(1)) \\
&= -\frac{\theta_0}{\theta - \theta_0}(1 + o(1)),
\end{aligned}$$

where only the first-order approximation is needed and $E[.]$ is taken with respect to Y. Similarly,

$$\frac{\partial}{\partial \theta} P_{\theta_0 \theta}(\tau_{-M} = \infty) = -\frac{1}{\theta_0} E\left[Y e^{\frac{\theta}{\theta_0}Y}\right](1 + o(1))$$

$$= -\frac{\theta_0}{(\theta - \theta_0)^2}(1 + o(1)).$$

On the other hand, from Theorem 2 in Wu (1999), we know that

$$E_{\theta_0}[\sigma_M P_\theta(\tau_{-M} = \infty)] = \frac{1}{2\theta_0^2} - \frac{1}{2(\theta - \theta_0)^2} + O(1).$$

A similar argument shows that

$$E_{\theta_0}[M P_\theta(\tau_{-M} = \infty)] = -\frac{1}{2\theta_0} E\left(Y\left(1 - e^{\frac{\theta}{\theta_0}Y}\right)\right)(1 + o(1))$$

$$= -\frac{1}{2\theta_0} + \frac{\theta_0}{2(\theta - \theta_0)^2}(1 + o(1)).$$

After some algebraic simplifications, we have the following local approximation.

Corollary 4.5: *As* $\theta_0, \theta \to 0$ *at the same order, and* $\theta d^{1-\epsilon} \to \infty$,

$$E^\nu[\hat{\theta} - \theta|N > \nu] = \frac{1}{d}\left(2 - \frac{\theta^3}{2(\theta - \theta_0)\theta_0^2}\right)(1 + o(1)).$$

At $\theta = -\theta_0$,

$$E^\nu[\hat{\theta} - \theta|N > \nu] = \frac{7}{4d}(1 + o(1). \tag{4.6}$$

4.2.2   A Corrected Normal Pivot

Now, we consider a corrected normal pivot based on

$$Z_N = \sqrt{N - \hat{\nu}}(T_N/(N - \hat{\nu}) - \theta)$$

by finding the first-order bias of Z_N and variance after bias correction. The asymptotic normality of Z_N can be established similar to Lemma 4.3 and the details are omitted.

We first write by using the total probability law

$$E^\nu[Z_N; N > \nu] = E_\theta[Z_{N_0}|S_{N_0} > d]P_{\theta_0 \theta}(\tau_{-M} < \infty)$$

$$+ E_{\theta_0 \theta}\left[\sqrt{N_M + \sigma_M}\left(\frac{S_{N_M}}{N_M + \sigma_M} - \theta\right); S_{N_M} > d\right],$$

and note that

$$E_{\theta_0 \theta}\left[\sqrt{N_M + \sigma_M}\left(\frac{S_{N_M}}{N_M + \sigma_M} - \theta\right); S_{N_M} > d\right]$$

$$= E_{\theta_0\theta} \left[\sqrt{N_M} \left(\frac{S_{N_M}}{N_M + \sigma_M} - \theta \right); S_{N_M} > d \right] (1 + o(1)).$$

By applying the same techniques in Theorems 4.2 and 4.4, we have the following result.

Theorem 4.5: *As $\nu \to \infty$ and d approaches ∞, for any given $\theta > 0$,*

$$E^\nu[Z_N | N > \nu] = \frac{b(\theta)}{\sqrt{d}} (1 + o(1)),$$

where

$$
\begin{aligned}
b(\theta) &= \theta^{-1/2} \Big(\frac{1}{2} + \frac{P_{\theta_0\theta}(\tau_{-M} < \infty)}{P_\theta(\tau_- = \infty)} \frac{\partial}{\partial\theta} P_\theta(\tau_- = \infty) + \frac{\partial}{\partial\theta} P_{\theta_0\theta}(\tau_{-M} = \infty) \\
&\quad + \theta E_{\theta_0}[(M - \theta\sigma_M) P_\theta(\tau_{-M} = \infty)] \Big).
\end{aligned}
$$

Similar to Corollary 4.5, we have the following local approximation.

Corollary 4.6: *Locally, as $\theta_0, \theta \to 0$ at the same order and $\theta d^{1-\epsilon} \to \infty$,*

$$E^\nu[Z_N | N > \nu] = \frac{1}{(\theta d)^{1/2}} \left(\frac{3}{2} - \frac{\theta^3}{2(\theta - \theta_0)\theta_0^2} \right) (1 + o(1)).$$

At $\theta = -\theta_0$,

$$E^\nu[Z_N | N > \nu] = \frac{5}{4(\theta d)^{1/2}} (1 + o(1)). \tag{4.7}$$

Finally, we evaluate the variance for the bias-corrected pivot. Similar to the proof for Theorem 4.3, we first write

$$E^\nu[(Z_N - \frac{b(\hat\theta)}{d^{1/2}})^2 | N > \nu) = E^\nu[Z_N^2 | N > \nu] - \frac{b^2(\theta)}{d} - \frac{2\theta^{1/2}}{d} b'(\theta). \tag{4.8}$$

For the first term of (4.8), we first write by using total probability law as

$$
\begin{aligned}
E^\nu[Z_N^2 | N > \nu] &= E[Z_{N_0}^2 | S_{N_0} > d] P_{\theta_0\theta}(\tau_{-M} < \infty) \\
&\quad + E_{\theta_0\theta} \left[(N_M + \sigma_M) \left(\frac{S_{N_M}}{N_M + \sigma_M} - \theta \right)^2; S_{N_M} > d \right].
\end{aligned}
$$

Now, by using the normal density property,

$$
\begin{aligned}
&E_{\theta_0\theta} \left[(N_M + \sigma_M) \left(\frac{S_{N_M}}{N_M + \sigma_M} - \theta \right)^2; S_{N_M} > d \right] \\
&= E_{\theta_0\theta} \left[N_M \left(\hat\theta_M - \theta + \frac{M - \theta\sigma_M}{N_M} \right)^2; S_{N_M} > d \right]
\end{aligned}
$$

$$+ \frac{\theta}{d} E_{\theta_0 \theta}[\sigma_M; \tau_{-M} = \infty](1 + o(1))$$

$$= E_{\theta_0 \theta}[N_M(\hat{\theta}_M - \theta)^2; S_{N_M} > d] + 2E_{\theta_0 \theta}\left[N_M(\hat{\theta}_M - \theta)\frac{M - \theta\sigma_M}{N_M}; S_{N_M} > d\right]$$

$$+ E_{\theta_0 \theta}\left[N_M\left(\frac{M - \theta\sigma_M}{N_M}\right)^2; S_{N_M} > d\right] + \frac{\theta}{d} E_{\theta_0 \theta}[\sigma_M; \tau_{-M} = \infty](1 + o(1))$$

$$= P_{\theta_0 \theta}(\tau_{-M} = \infty) + \frac{1}{d}\frac{\partial^2}{\partial^2 \theta}(\theta P_{\theta_0 \theta}(\tau_{-M} = \infty))$$

$$+ \frac{2}{d} E_{\theta_0}\left[\frac{\partial}{\partial \theta}((\theta M - \theta^2 \sigma_M) P_\theta(\tau_{-M} = \infty))\right]$$

$$+ \frac{\theta}{d} E_{\theta_0}([(\theta\sigma_M - M)^2 + \sigma_M] P_\theta(\tau_{-M} = \infty))(1 + o(1)).$$

Combining the above approximations, we have the following result.

Theorem 4.6: *As $\nu, d \to \infty$, for any $\theta > 0$,*

$$E^\nu\left[\left(Z_N - \frac{b(\hat{\theta})}{d^{1/2}}\right)^2 | N > \nu\right] = 1 + \frac{c(\theta)}{d}(1 + o(1)),$$

where

$$c(\theta) = \frac{P_{\theta_0 \theta}(\tau_{-M} < \infty)}{P_\theta(\tau_- = \infty)}\frac{\partial^2}{\partial^2 \theta}(\theta P_\theta(\tau_- = \infty)) + \frac{\partial^2}{\partial^2 \theta}(\theta P(\tau_{-M} = \infty))$$

$$+ 2\frac{\partial}{\partial \theta} E_{\theta_0}[(\theta M - \theta^2 \sigma_M) P_\theta(\tau_{-M} = \infty)]$$

$$+ \theta E_{\theta_0}[((\theta\sigma_M - M)^2 + \sigma_M) P(\tau_{-M} = \infty)] - b^2(\theta) - 2\theta^{1/2} b'(\theta).$$

A local expansion for the second moment is technically complicated and the details are given in Section 4.5.2.

Corollary 4.7: *As $\theta_0, \theta \to 0$ at the same order and $\theta d^{1-\epsilon} \to \infty$,*

$$E^\nu[(Z_N - \frac{b(\hat{\theta})}{d^{1/2}})^2 | N > \nu]$$

$$= 1 + \frac{1}{d}\{\frac{\theta_0^2 - 2\theta^2 - \theta\theta_0}{(\theta - \theta_0)^3} - \frac{2\theta_0}{\theta(\theta - \theta_0)}$$

$$- \frac{2\theta + \theta_0}{\theta_0^2} + \frac{\theta(\theta + \theta_0)^2}{2\theta_0^2}\left(\frac{1}{\theta_0^2} + \frac{\theta_0}{(\theta - \theta_0)^3}\right)$$

$$+ \frac{\theta}{2}\left(\frac{\theta^2}{\theta_0^2} + 1\right)\left(\frac{1}{\theta_0^2} - \frac{1}{(\theta - \theta_0)^2}\right) - \frac{1}{\theta}\left(\frac{3}{2} - \frac{\theta^3}{2(\theta - \theta_0)\theta_0^2}\right)^2$$

$$+ \frac{3}{2\theta} + \frac{\theta^2}{\theta_0^2}\left(\frac{5}{2(\theta - \theta_0)} - \frac{\theta}{(\theta - \theta_0)^2}\right)\}(1 + o(1)).$$

At $\theta = -\theta_0$,

$$E^\nu[(Z_N - b(\hat\theta)/d^{1/2})^2 | N > \nu] = 1 + \frac{27/16}{d\theta}(1 + o(1)). \qquad (4.9)$$

Thus, a corrected pivot can be formed as

$$Z_N^* = \left(Z_N - \frac{b(\hat\theta)}{d^{1/2}}\right)\left(1 + \frac{1}{d}c(\hat\theta)\right)^{-1/2},$$

which is asymptotically normal up to the order $o(1/d)$. An estimator for θ can be defined as

$$\theta^* = \frac{T_N}{N - \hat\nu} - \frac{b(\hat\theta)}{d^{1/2}(N - \hat\nu)^{1/2}},$$

which is an approximate median and has the average bias correction of

$$\frac{1}{d}\left(\frac{3}{2} - \frac{\theta^3}{2(\theta - \theta_0)\theta_0^2}\right).$$

A $(1 - \alpha)$-level confidence interval for θ can be formed as

$$\hat\theta - \frac{b(\hat\theta)}{d^{1/2}(N - \hat\nu)^{1/2}} \pm z_{\alpha/2}\frac{(1 + c(\hat\theta)/d)^{1/2}}{(N - \hat\nu)^{1/2}}.$$

Similarly, a bias-corrected estimation $\tilde\theta$ can be obtained by letting

$$\tilde\theta = \hat\theta - \frac{a(\hat\theta)/d}{1 + a'(\hat\theta)/d},$$

which has a local bias correction

$$\frac{1}{d}\left(2 - \frac{\theta^3}{2(\theta - \theta_0)\theta_0^2}\right).$$

4.3   Numerical Examples and Discussions

In this section, we conduct a simulation study to check the results and the accuracy of the approximations.

Table 4.1 gives the comparison between the simulated and approximated bias for $\hat\theta_0$ and

$$Z_{N_0} = \sqrt{N_0}(S_{N_0}/N_0 - \theta)$$

from Section 4.2. Here we use $E^*[.]$ and $P^*(.)$ to denote the conditional version $E[.|S_{N_0} > d]$ and $P[.|S_{N_0} > d]$. For a very typical value $d = 10$, and θ takes values 0.1, 0.25, and 0.5, 10,000 simulations are conducted and the simulated values for $P_\theta(S_{N_0} > d)$ are also reported. The conditional probability and expectation are estimated based on those simulations with $S_{N_0} > d$. Approximate

Table 4.1: Bias and Coverage Probability with $d = 10$ in Section 2

θ		$P_\theta(S_{N_0} > d)$	$E^*[\hat{\theta}_0 - \theta]$	$E^*[Z_{N_0}]$	$P^*(|Z^*_{N_0}| < 1.96)$
0.1	sim.	0.145	0.276 (0.003)	1.317 (0.013)	0.922
	app.		0.194	1.442	0.950
0.25	sim.	0.295	0.213 (0.003)	0.856 (0.011)	0.948
	app.		0.185	0.860	0.950
0.5	sim.	0.530	0.173 (0.003)	0.521 (0.012)	0.958
	app.		0.171	0.541	0.950

values are calculated based on (4.1) and (4.2) from Corollary 4.1 and Corollary 4.3. Listed also are the simulated coverage probability $P^*(|Z^*_{N_0}| < 1.96)$ based on (4.3) given in Corollary 4.4 where we use the approximations for the mean and variance:

$$Z^*_{N_0} = (Z_{N_0} - (3/2 - \hat{\theta}_0\rho)/(d\hat{\theta}_0)^{1/2})/(1 + 5/(4d\hat{\theta}_0))^{1/2}.$$

The standard errors are reported in the parentheses for the bias estimations.

From Table 4.1, we see that the approximations perform quite satisfactorily, particularly for $\theta = 0.25$. The approximation for the bias of $\hat{\theta}_0$ is slightly low. The approximation for the mean of Z_{N_0} is very accurate, which makes the coverage probability approximation satisfactory for $\theta = 0.25$ and 0.5.

In Table 4.2, we conduct a similar comparison for the post-change mean estimator considered in Section 4.3. For $-\theta_0 = \theta = 0.1, 0.25$, and 0.5, and change point $\nu = 0, 25$, and 50 with $d = 10$, 5000 simulations are conducted. Conditioning on $N > \nu$ (the simulated probabilities are reported), simulated biases of $\hat{\theta}$ and

$$Z_N = \sqrt{T - \hat{\nu}}(\hat{\theta} - \theta)$$

are reported along with the approximated values based on (4.6) and (4.7) from Corollaries 4.5 and 4.6. Standard errors are reported in the parentheses. Again, we use $E^*[.]$ to denote the conditional expectation $E[.|N > \nu]$. The conditional probability and expectations are estimated based on those simulations with $N > \nu$.

Due to the complicated form of Corollary 4.7, we simulated the coverage probability for the corrected confidence interval based on the theoretical corrected pivot. That means, we correct the bias of Z_N with the theoretical bias $b(\theta)/d^{1/2}$, which is equal to $5/(4(d\theta)^{1/2})$ locally, while the variance $E^\nu[(Z_N - b(\theta)/d^{1/2})^2|N > \nu]$ of Z_N is approximated as

$$1 - 13/(16d\theta)$$

since the term of $-2\theta^{1/2}b'(\theta)/d$ [which is equal to $5/(2d\theta)$ locally from last equation in Section 4.2.2] no longer exists in Eq. (4.8). Thus, we use

$$Z^*_N = (Z_N - 5/(4(d\theta)^{1/2}))/(1 - 13/(16d\theta))^{1/2},$$

Table 4.2: Biases and Coverage Probability with $d = 10$ in Section 4.3

$-\theta_0 = \theta$	ν	$P_{\theta_0}(N > \nu)$	$E^*[\hat{\theta} - \theta]$	$E^*[Z_N]$	$P^*(|Z_N^*| < 1.96)$
0.1	0	1.000	0.279(0.002)	1.342(0.013)	0.779
	25	0.984	0.241(0.003)	1.267(0.013)	0.801
	50	0.910	0.232(0.003)	1.168(0.013)	0.805
	app.		0.175	1.25	0.950
0.25	0	1.000	0.214(0.003)	0.846(0.012)	0.942
	25	0.996	0.176(0.003)	0.702(0.011)	0.940
	50	0.987	0.172(0.003)	0.674(0.012)	0.938
	app.		0.175	0.791	0.950
0.5	0	1.000	0.173(0.004)	0.537(0.014)	0.948
	25	1.000	0.144(0.004)	0.415(0.013)	0.942
	50	1.000	0.142(0.004)	0.411(0.013)	0.936
	app.		0.175	0.559	0.950

where the approximation for the variance is based on Eq. (4.9).

From Table 4.2, we find several interesting facts. First, the convergence of the biases is extremely fast as ν gets larger. In fact, we see that the results for $\nu = 25$ and 50 have very little differences. Second, for the bias of $\hat{\theta}$, the approximate values at $\theta = 0.25$ are quite accurate. Third, the approximate values for the coverage probability are very accurate for $\theta = 0.25$ and 0.5, while they are less satisfactory for $\theta = 0.1$. The reasons is that the approximation is obtained by first assuming $d \to \infty$ and then letting $\theta \to 0$. Thus, when $d\theta$ is not very large, a poor approximation is expected. A moderate deviation analysis by assuming $d\theta \to constant$ should improve the accuracy of the approximations and will be a future topic.

4.4   Proofs

4.4.1   Proof of Lemma 4.2

For any given $\epsilon > 0$, from the total probability law, we can write

$$P_\theta \left(\left| \frac{N_0}{d/\theta} - 1 \right| > \epsilon; S_{N_0} > d \right) \leq P_\theta \left(\left| \frac{\tau_d}{d/\theta} - 1 \right| > \epsilon \right).$$

From the strong law of large numbers, we have

$$P_\theta \left(\left| \frac{\tau_d}{d/\theta} - 1 \right| > \epsilon \right) \to 0,$$

as $d \to \infty$. Combining with the result from Lemma 4.2, we get the first result.

For the second result, by using Wald's likelihood ratio identity, we have

$$P_\theta(S_{N_0} > d) = E_0\left[\exp\left(\theta S_{N_0} - \frac{\theta^2}{2}N\right); S_{N_0} > d\right]. \qquad (4.10)$$

By differentiating (4.10) once with respect to θ, we get

$$\frac{\partial}{\partial\theta}P_\theta(S_{N_0} > d) = E_\theta[S_{N_0}; S_{N_0} > d] - \theta E_\theta[N_0; S_{N_0} > d].$$

Thus,

$$E_\theta[N_0; S_{N_0} > d] = \frac{1}{\theta}\left[E_\theta[S_{N_0}; S_{N_0} > d] - \frac{\partial}{\partial\theta}P_\theta(S_{N_0} > d)\right].$$

Since

$$\begin{aligned}
E_\theta[S_{N_0}|S_{N_0} > d] &= d + E_\theta[S_{N_0} - d|S_{N_0} > d] \\
&= d + E_\theta R_\infty + o(1),
\end{aligned}$$

and by using Wald's identity,

$$\begin{aligned}
\frac{\partial}{\partial\theta}P_\theta(S_{N_0} > d) &= E_\theta[(S_{N_0} - \theta N_0); S_{N_0} > d] \\
&= E_\theta[S_{\tau_d} - \theta\tau_d] - E_\theta[S_{\tau_d} - \theta\tau_d; S_{N_0} \le 0] \\
&= -E_\theta[E_\theta[S_{\tau_d} - \theta\tau_d|S_{N_0}]; S_{N_0} \le 0] \\
&= -E_\theta[S_{N_0} - \theta N_0; S_{N_0} \le 0] \\
&= -E_\theta[S_{\tau_-} - \theta\tau_-; \tau_- < \infty] + o(1) \\
&= \frac{\partial}{\partial\theta}P_\theta(\tau_- = \infty) + o(1),
\end{aligned}$$

where in the third equation from the last, we note that given $S_{N_0} = x \le 0$ and $N_0 = k$, $S_{n+k} - x$ for $n \ge 0$ behaves like a random walk with boundary $d - x$, and τ_d behaves like $k + \tau_{d-x}$ and thus, by using Wald's identity again, we have

$$\begin{aligned}
E_\theta[S_{\tau_d} - \theta\tau_d|S_{N_0} = x; N_0 = k] &= E_\theta[S_{\tau_{d-x}} + x - \theta(k + \tau_{d-x})] \\
&= x - \theta k.
\end{aligned}$$

Thus, the second result is proved.

For the variance of N_0 given $S_{N_0} > d$, by differentiating (4.10) twice with respect to θ, we get

$$\frac{\partial^2}{\partial^2\theta}P_\theta(S_{N_0} > d) = -E_\theta[N_0; S_{N_0} > d] + E_\theta[(\theta N_0 - S_{N_0})^2; S_{N_0} > d].$$

Thus,

$$\begin{aligned}
E_\theta[N_0^2; S_{N_0} > d] &= \frac{1}{\theta^2}[2E_\theta[N_0 S_{N_0}; S_{N_0} > d] - E_\theta[S_{N_0}^2; S_{N_0} > d] \\
&\quad + E_\theta[N_0; S_{N_0} > d] + \frac{\partial^2}{\partial^2\theta}P_\theta(S_{N_0} > d)].
\end{aligned}$$

Next, by differentiating the following equation with respect to θ once,

$$E_\theta[S_{N_0}; S_{N_0} > d] = E_0\left[S_{N_0}\exp(\theta S_{N_0} - \frac{\theta^2}{2}N_0); S_{N_0} > d\right],$$

we get

$$E_\theta[N_0 S_{N_0}; S_{N_0} > d] = \frac{1}{\theta}\left[E_\theta[S_{N_0}^2; S_{N_0} > d] - \frac{\partial}{\partial\theta}E_\theta[S_{N_0}; S_{N_0} > d]\right].$$

Thus,

$$\begin{aligned}E_\theta[N_0^2; S_{N_0} > d] &= \frac{1}{\theta^2}[E_\theta[S_{N_0}^2; S_{N_0} > d] - 2\frac{\partial}{\partial\theta}E_\theta[S_{N_0}; S_{N_0} > d]\\ &\quad + E_\theta[N_0; S_{N_0} > d] + \frac{\partial^2}{\partial^2\theta}P_\theta(S_{N_0} > d)].\end{aligned}$$

By denoting $S_{N_0} - d = R_{N_0}$ given $S_{N_0} > d$, we have

$$E_\theta[S_{N_0}^2|S_{N_0} > d] = d^2 + 2dE_\theta[R_{N_0}|S_{N_0} > d] + E_\theta[R_{N_0}^2|S_{N_0} > d]$$

and

$$\frac{\partial}{\partial\theta}E_\theta[S_{N_0}; S_{N_0} > d] = d\frac{\partial}{\partial\theta}P_\theta(S_{N_0} > d) + \frac{\partial}{\partial\theta}E_\theta[R_{N_0}; S_{N_0} > d].$$

Thus,

$$\begin{aligned}E_\theta[N_0^2|S_{N_0}] &= \frac{1}{\theta^2}[d^2 + 2dE_\theta[R_{N_0}|S_{N_0} > d]\\ &\quad - 2d\frac{\frac{\partial}{\partial\theta}P_\theta(S_{N_0} > d)}{P_\theta(S_{N_0} > d)} + \frac{d}{\theta} + O(1)],\end{aligned} \qquad (4.11)$$

where, similar to the proof for the second result, we can show that the $O(1)$ term exists and is finite as $d \to \infty$.

On the other hand, from the second result, we have

$$\begin{aligned}(E_\theta[N_0|S_{N_0} > d])^2 &= \frac{1}{\theta^2}[d^2 + 2dE_\theta[R_{N_0}|S_{N_0} > d]\\ &\quad - 2d\frac{\frac{\partial}{\partial\theta}P_\theta(S_{N_0} > d)}{P_\theta(S_{N_0} > d)} + O(1)].\end{aligned} \qquad (4.12)$$

By combining (4.11) and (4.12) up to the order $O(1)$, we find

$$\begin{aligned}\text{Var}_\theta(N_0|S_{N_0} > d) &= E_\theta[N_0^2|S_{N_0} > d] - (E_\theta[N_0|S_{N_0}])^2\\ &= \frac{d}{\theta^3} + O(1).\end{aligned}$$

A detailed analysis continuing the above lines by following the discussion in Section 4.1 for the sample mean can also give the second-order expansions for the first two moments of N_0 given $S_{N_0} > d$ and thus can be used to establish the corrected normal approximations for N_0 as well.

4.4.2 Derivation of Corollary 4.7

We first derive the joint moment-generating function of M and σ_M, which is of independent interest. Although the results are quite standard, we did not find a convenient reference. Write

$$\sigma_M = \sum_{i=1}^{K} \tau_+^{(i)}; \quad M = \sum_{i=1}^{K} S_{\tau_+^{(i)}},$$

where $\{(\tau_+^{(i)}, S_{\tau_+^{(i)}})\}$ are i.i.d. copies of (τ_+, S_{τ_+}) and

$$K = \sup\{k > 0 : \tau_+^{(k)} < \infty\},$$

which is a geometric random variables with terminating probability

$$p = P_{\theta_0}(\tau_+ = \infty)$$

and

$$P(K = k) = (1-p)^{k-1}p, \quad \text{for} \quad k \geq 1.$$

Thus, the joint moment-generating function of M and σ_M can be calculated as

$$E_{\theta_0}\left[e^{\lambda\sigma_M + tM}\right] = \sum_{k=0}^{\infty} E_{\theta_0}\left[e^{\lambda\sum_{i=1}^{k}\tau_+^{(i)} + t\sum_{i=1}^{k}S_{\tau_+^{(i)}}}; K = k\right]$$

$$= \sum_{k=0}^{\infty} E_{\theta_0}\left[e^{\lambda\sum_{i=1}^{k}\tau_+^{(i)} + t\sum_{i=1}^{k}S_{\tau_+^{(i)}}}; \tau_+^{(1)} < \infty; ...; \tau_+^{(k)} < \infty; \tau_+^{(k+1)} = \infty\right]$$

$$= \sum_{k=0}^{\infty} \left(E_{\theta_0}\left[e^{\lambda\tau_+ + tS_{\tau_+}}; \tau_+ < \infty\right]\right)^k P_{\theta_0}(\tau_+ = \infty)$$

$$= \frac{P_{\theta_0}(\tau_+ = \infty)}{1 - E_{\theta_0}\left[e^{\lambda\tau_+ + tS_{\tau_+}}; \tau_+ < \infty\right]}.$$

Thus,

$$E_{\theta_0}\left[(\theta\sigma_M - M)^2 e^{-2\theta M}\right]$$

$$= P_{\theta_0}(\tau_+ = \infty)$$

$$\times \left\{2\frac{\left(E_{\theta_0}\left(S_{\tau_+}e^{-2\theta S_{\tau_+}}; \tau_+ < \infty\right) - \theta E_{\theta_0}\left(\tau_+ e^{-2\theta S_{\tau_+}}; \tau_+ < \infty\right)\right)^2}{\left(1 - E_{\theta_0}\left(e^{-2\theta S_{\tau_+}}; \tau_+ < \infty\right)\right)^3}\right.$$

$$\left. + \frac{E_{\theta_0}\left[(\theta^2\tau_+ - 2\theta\tau_+ S_{\tau_+} + S_{\tau_+}^2)e^{-2\theta S_{\tau_+}}; \tau_+ < \infty\right]}{\left(1 - E_{\theta_0}\left(e^{-2\theta S_{\tau_+}}; \tau_+ < \infty\right)\right)^2}\right\}.$$

As $\theta_0, \theta \to 0$ at the same order, from Lemma 11 of Wu (1999),

$$E_{\theta_0}\left(\tau_+^2 e^{-2\theta S_{\tau_+}}; \tau_+ < \infty\right) = -\frac{E_0 S_{\tau_+}}{\theta_0^3}(1 + o(1)).$$

Also,

$$E_{\theta_0}\left(\tau_+ e^{-2\theta S_{\tau_+}}; \tau_+ < \infty\right) = -\frac{E_0 S_{\tau_+}}{\theta_0}(1 + o(1)),$$

$$E_{\theta_0}\left(S_{\tau_+}^k e^{-2\theta S_{\tau_+}}; \tau_+ < \infty\right) = E_0 S_{\tau_+}^k (1 + o(1)), \quad \text{for } k = 1, 2, \ldots,$$

$$P_{\theta_0}(\tau_+ = \infty) = -2\theta_0 E_0 S_{\tau_+}(1 + o(1)),$$

$$1 - E_{\theta_0}\left(e^{-2\theta S_{\tau_+}}; \tau_+ < \infty\right) = 2(\theta - \theta_0)E_0 S_{\tau_+}(1 + o(1)),$$

$$E_{\theta_0}\left(\tau_+ S_{\tau_+} e^{-2\theta S_{\tau_+}}; \tau_+ < \infty\right) = -\frac{E_0 S_{\tau_+}^2}{2\theta_0}(1 + o(1)),$$

where the last equation is obtained from Lemma 10.27 of Siegmund (1985). Thus, by ignoring the higher-order terms, we have

$$\begin{aligned}
E_{\theta_0}[(\theta \sigma_M - M)^2 P_\theta(\tau_{-M} < \infty)] &= E_{\theta_0}\left[(\theta \sigma_M - M)^2 e^{-2\theta M}\right](1 + o(1)) \\
&= -\frac{\theta_0(1 + \theta/\theta_0)^2}{2(\theta - \theta_0)^3} + \frac{\theta^2}{2\theta_0^2(\theta - \theta_0)^2}(1 + o(1)).
\end{aligned}$$

A simpler derivation gives

$$E_{\theta_0}[(\theta \sigma_M - M)^2] = \frac{(1 + \theta/\theta_0)^2}{2\theta_0^2} + \frac{\theta^2}{2\theta_0^4}(1 + o(1)).$$

For the other terms in Theorem 4.6, we have from the proof for Corollary 4.5,

$$\frac{\partial^2}{\partial^2 \theta}(\theta P_\theta(\tau_- = \infty))/P_\theta(\tau_- = \infty) = \frac{2}{\theta}(1 + o(1)),$$

$$\begin{aligned}
P_{\theta_0 \theta}(\tau_{-M} < \infty) &= E e^{\frac{\theta}{\theta_0} Y}(1 + o(1)) \\
&= -\frac{\theta_0}{\theta - \theta_0}(1 + o(1)),
\end{aligned}$$

$$\begin{aligned}
\frac{\partial^2}{\partial^2 \theta}(\theta P_{\theta_0 \theta}(\tau_{-M} = \infty)) &= \frac{\partial^2}{\partial^2 \theta}\left(\theta\left(1 - E_{\theta_0} e^{\frac{\theta}{\theta_0} Y}\right)\right)(1 + o(1)) \\
&= \frac{2\theta_0^2}{(\theta - \theta_0)^3}(1 + o(1)),
\end{aligned}$$

$$\frac{\partial}{\partial \theta} E_{\theta_0}[(\theta^2 \sigma_M - \theta M) P_\theta(\tau_{-M} = \infty)] = \frac{2\theta + \theta_0}{2\theta_0^2} + \frac{2\theta^2 + \theta\theta_0 + \theta_0^2}{2(\theta - \theta_0)^3},$$

$$E_{\theta_0}[\sigma_M P_\theta(\tau_{-M} = \infty)] = \frac{1}{2\theta_0^2} - \frac{1}{2(\theta - \theta_0)^2}(1 + o(1)),$$

and, finally,

$$-2\theta^{1/2} b'(\theta) = \frac{3}{2\theta} + \frac{\theta^2}{\theta_0^2}\left(\frac{5}{2(\theta - \theta_0)} - \frac{\theta}{(\theta - \theta_0)^2}\right).$$

Table 4.3: Post-Change Mean Estimator for Nile River

d	$N - \hat{\nu}$	Post-mean estimator $\hat{\theta}$	Bias $\hat{\theta} - 1$	Theo. approx.
5	4	1.436	0.436	0.350
10	9	1.343	0.343	0.175
15	15	1.231	0.231	0.116
20	17	1.286	0.286	0.088
25	23	1.105	0.105	0.070
30	27	1.129	0.129	0.058
35	32	1.125	0.125	0.050
40	38	1.060	0.060	0.044
45	42	1.077	0.077	0.039
50	45	1.115	0.115	0.035
55	51	1.091	0.091	0.032
60	54	1.111	0.111	0.029
65	68	0.964	$-$ 0.036	0.027
70	71	0.988	$-$ 0.012	0.025

Combining the above approximations, we get the expected result after some simplifications.

4.5 Case Study

4.5.1 Nile River

Here we use the Nile River data set listed in Chapter 2 as an illustration for the post-change mean estimation.

We directly use the CUSUM process formed by the standardized data with $\theta = -\theta_0 = 1$.

The following table lists the sample size after the change-point estimator $\hat{\nu} = 28$ and the post-change mean estimator as well as the real biases and theoretical biases for the threshold value d changing from 5 to 70.

We see that the bias becomes smaller as d gets larger. The theoretical approximation $7/(4d)$ is slightly smaller than the observed value due to the large values of θ_0 and θ and also the small sample size, which causes relatively large errors.

This case shows that even in this small data set, the theoretical result shows the right scale for the bias correction.

4.5.2 Coal Mining Disaster

Although we have not given the theoretical results for the general exponential family case, we can give the post-change mean estimator without bias correction.

Table 4.4: Post-Change Mean Estimators for Coal Mining Disaster

Threshold (d)	Post-change sample size	Post-change sample mean
5	8	332.5
10	11	422.0
15	23	337.4
20	29	367.9
25	32	386.3
30	33	385.1
35	58	332.0
40	63	331.4

We use the same standardized CUSUM process as given in Chapter 2. The change-point estimator is always $\hat{\nu} = 46$ no matter what the threshold is (at least 5 and at most 40).

The following table gives the post-change sample size and the post-change sample mean (estimator for $1/\lambda$) for several values of d. The true value for the post-change mean is suppose to be 335.

We see that the bias is mostly positive and gets smaller as d gets larger. The theoretical study for the bias of $\hat{\lambda}$ will be an immediate topic.

5

Estimation After False Signal

In this chapter, we consider the asymptotic distribution of $\hat{\theta}$ if $T < \nu$ where ν is significantly large enough, that means, a false signal is made. By looking at the definition of $\hat{\theta}$ and $\hat{\nu}$, we see that as $\nu \to \infty$, $\hat{\theta}$ is indeed the sample mean for a conditional random walk staying positive until it crosses the boundary d.

In practical situations, we rarely know if a signal is true or false. A true signal is caused by a true changed segments; while a false signal is raised by a regular cluster without change. It is fundamental to distinguish whether a signal is true or false. The distribution of the post-change mean estimator can play a role in such a classification, which can be thought of as a mixture of two asymptotic normal distributions: one under the true signal and the other under the false signal. If the mean and variance can be determined, respectively, a classification rule can be defined correspondingly.

5.1 Conditional Random Walk with Negative Drift

The following notations are used in our discussion. Define

$$S_n = S_0 + \sum_{i=1}^{n} X_i.$$

For $S_0 = 0$, define

$$N_0 = \inf\{n > 0 : S_n < 0 \quad \text{or} \quad > d\}$$

and

$$\tau_d = \inf\{n > 0 : S_n > d\},$$

and denote by

$$R_d = S_{\tau_d} - d \quad \text{and} \quad R_\infty = \lim_{d \to \infty} R_d$$

if it exists. Furthermore, we define

$$\tau_- = \inf\{n > 0 : S_n \le 0\}; \quad \text{and} \quad \tau_+ = \inf\{n > 0; S_n > 0\}.$$

Thus, if the signal is false, then

$$\hat{\theta} = \frac{S_{N_0}}{N_0},$$

given $S_{N_0} > d$.

Throughout our discussion, we denote by $E_\theta[.]$ the expectation when the mean of the random walk is θ, and $E_\theta[X; A] = E_\theta[X I_A]$ and as a convention

$$E_\theta[X|A] = E_\theta[X; A]/P_\theta(A).$$

We first study the mean of $\hat{\theta} = S_{N_0}/N_0$ given $S_{N_0} > d$. Our analysis is split into two steps. In the first step, we first assume $d \to \infty$ and obtain the first-order asymptotic results. Then in the second step, we shall give the local approximation for the asymptotic results by further letting $\theta_0 \to 0$. Detailed verifications are omitted. Readers are referred to Chapter 4 in the true signal case.

By using normal density property and Wald's likelihood ratio identity [e.g. Siegmund (1985, Prop. 2.24)], we have

$$E_{\theta_0}\left[\frac{S_{N_0}}{N_0} - \theta_0 | S_{N_0} > d\right] = \frac{E_{\theta_0}\left[\frac{S_{N_0} - \theta_0 N_0}{N_0}; S_{N_0} > d\right]}{P_{\theta_0}(S_{N_0} > d)}$$

$$= \frac{\frac{\partial}{\partial \theta_0} E_{\theta_0}\left[\frac{1}{N_0}; S_{N_0} > d\right]}{P_{\theta_0}(S_{N_0} > d)}$$

$$= \frac{\frac{\partial}{\partial \theta_0} E_{-\theta_0}\left[\frac{1}{N_0} e^{2\theta_0 S_{N_0}}; S_{N_0} > d\right]}{E_{-\theta_0}\left[e^{2\theta_0 S_{N_0}}; S_{N_0} > d\right]}$$

$$= \frac{\frac{\partial}{\partial \theta_0}\left[e^{2\theta_0 d} E_{-\theta_0}\left[\frac{1}{N_0} e^{2\theta_0(S_{N_0} - d)}; S_{N_0} > d\right]\right]}{e^{2\theta_0 d} E_{-\theta_0}\left[e^{2\theta_0(S_{N_0} - d)}; S_{N_0} > d\right]}$$

$$= \frac{2d E_{-\theta_0}\left[\frac{1}{N_0} e^{2\theta_0(S_{N_0} - d)}; S_{N_0} > d\right]}{E_{-\theta_0}\left[e^{2\theta_0(S_{N_0} - d)}; S_{N_0} > d\right]} + \frac{\frac{\partial}{\partial \theta_0} E_{-\theta_0}\left[\frac{1}{N_0} e^{2\theta_0(S_{N_0} - d)}; S_{N_0} > d\right]}{E_{-\theta_0}\left[e^{2\theta_0(S_{N_0} - d)}; S_{N_0} > d\right]}$$

$$= 2dE_{\theta_0}\left[\frac{1}{N_0}|S_{N_0} > d\right] + \frac{\frac{\partial}{\partial\theta_0}E_{-\theta_0}\left[\frac{1}{N_0}e^{2\theta_0(S_{N_0}-d)}; S_{N_0} > d\right]}{E_{-\theta_0}\left[e^{2\theta_0(S_{N_0}-d)}; S_{N_0} > d\right]}.$$

For the second term in the above equation, we have the following lemma, which is given in Chapter 3.

Lemma 5.1: *Under* $P_{-\theta_0}(.)$, *as* $d \to \infty$, *for any fixed* θ_0,

$$N_0 = -\frac{d}{\theta_0}(1 + o_p(1)),$$

$$E_{-\theta_0}[N_0|S_{N_0} > d] = -\frac{d}{\theta_0}(1 + o(1)),$$

$$Var_{-\theta_0}(N_0|S_{N_0} > d) = -\frac{d}{\theta_0^3}(1 + o(1)).$$

As $d \to \infty$, from the derivation of (10.19) of Siegmundv(1985), we have

$$E_{-\theta_0}\left[e^{2\theta_0(S_{N_0}-d)}; S_{N_0} > d\right] = E_{-\theta_0}e^{2\theta_0 R_\infty}P_{-\theta_0}(\tau_- = \infty)(1 + o(1)).$$

Thus, similar to the proof for Theorem 4.1, we have

$$\frac{\frac{\partial}{\partial\theta_0}E_{-\theta_0}\left[\frac{1}{N_0}e^{2\theta_0(S_{N_0}-d)}; S_{N_0}>d\right]}{E_{-\theta_0}\left[e^{2\theta_0(S_{N_0}-d)}; S_{N_0}>d\right]}$$

$$= -\frac{1}{d}\frac{\frac{\partial}{\partial\theta_0}(\theta_0 E_{-\theta_0}e^{2\theta_0 R_\infty}P_{-\theta_0}(\tau_- = \infty))}{E_{-\theta_0}e^{2\theta_0 R_\infty}P_{-\theta_0}(\tau_- = \infty)}(1 + o(1)).$$

Thus,

$$E_{\theta_0}\left(\frac{S_{N_0}}{N_0} + \theta_0|S_{N_0} > d\right) = 2dE_{\theta_0}\left[\left(\frac{1}{N_0} + \frac{\theta_0}{d}\right)|S_{N_0} > d\right]$$

$$-\frac{1}{d}\frac{\frac{\partial}{\partial\theta_0}(\theta_0 E_{-\theta_0}e^{2\theta_0 R_\infty}P_{-\theta_0}(\tau_- = \infty))}{E_{-\theta_0}e^{2\theta_0 R_\infty}P_{-\theta_0}(\tau_- = \infty)}(1 + o(1)). \qquad (5.1)$$

On the other hand, we can write

$$E_{\theta_0}\left(\frac{S_{N_0}}{N_0} + \theta_0|S_{N_0} > d\right) = E_{\theta_0}\left[\frac{S_{N_0} - d}{N_0}|S_{N_0} > d\right]$$

$$+ dE_{\theta_0}\left[\left(\frac{1}{N_0} + \frac{\theta_0}{d}\right)|S_{N_0} > d\right]. \qquad (5.2)$$

By subtracting Eq. (5.1) from twice of Eq. (5.2), we get

$$E_{\theta_0}\left(\frac{S_{N_0}}{N_0} + \theta_0|S_{N_0} > d\right) = 2E_{\theta_0}\left[\frac{S_{N_0}-d}{N_0}|S_{N_0} > d\right]$$

$$+\frac{1}{d}\frac{\frac{\partial}{\partial\theta_0}\left(\theta_0 E_{-\theta_0}e^{2\theta_0 R_\infty}P_{-\theta_0}(\tau_- = \infty)\right)}{E_{-\theta_0}e^{2\theta_0 R_\infty}P_{-\theta_0}(\tau_- = \infty)}(1 + o(1)). \qquad (5.3)$$

Note that

$$2E_{\theta_0}\left[\frac{S_{N_0}-d}{N_0}\Big|S_{N_0}>d\right] = -\frac{2\theta_0}{d}E_{\theta_0}[S_{N_0}-d|S_{N_0}>d](1+o(1))$$

$$= -\frac{2\theta_0}{d}\frac{E_{-\theta_0}(R_\infty e^{2\theta_0 R_\infty})}{E_{-\theta_0}e^{2\theta_0 R_\infty}}(1+o(1)).$$

Thus, we have the following

Theorem 5.1: *As* $d \to \infty$,

$$E_{\theta_0}\left(\frac{S_{N_0}}{N_0}+\theta_0\Big|S_{N_0}>d\right) = \frac{1}{d}\{\frac{\frac{\partial}{\partial\theta_0}(\theta_0 E_{-\theta_0}e^{2\theta_0 R_\infty}P_{-\theta_0}(\tau_-=\infty))}{E_{-\theta_0}e^{2\theta_0 R_\infty}P_{-\theta_0}(\tau_-=\infty)}$$

$$- 2\theta_0\frac{E_{-\theta_0}R_\infty e^{2\theta_0 R_\infty}}{E_{-\theta_0}e^{2\theta_0 R_\infty}}\}(1+o(1)). \quad (5.4)$$

The theorem shows that the mean for the conditional random walk $\{S_n\}$ given $S_{N_0} > d$ is asymptotically $-\theta_0$.

As a byproduct, by subtracting (5.2) from (5.1), we get

$$E_{\theta_0}\left[\left(\frac{1}{N_0}+\frac{\theta_0}{d}\right);S_{N_0}>d\right] = \frac{1}{d^2}\{\frac{\partial}{\partial\theta_0}(\theta_0 E_{-\theta_0}e^{2\theta_0 R_\infty}P_{-\theta_0}(\tau_-=\infty))$$

$$-\theta_0 E_{\theta_0}(R_\infty e^{2\theta_0 R_\infty})P_{-\theta_0}(\tau_-=\infty)\}(1+o(1)), \quad (5.5)$$

which will be used in the next section.

In the following, we give a local approximation by further assuming that $\theta_0 \to 0$. Detailed techniques can be seen in Chapter 1.3.

We first note that [for example, Corollary 8.39 of Siegmund (1985)]

$$P_{-\theta_0}(\tau_-=\infty) = \frac{1}{E_{-\theta_0}\tau_+}$$

$$= -\frac{\theta_0}{E_{-\theta_0}S_{\tau_+}}$$

$$= -\frac{\theta_0}{E_0 S_{\tau_+}}e^{\theta_0\rho}(1+o(\theta_0^2)),$$

where the last equation is obtained from Lemma 10.27 of Siegmund (1985) and $\rho \approx 0.583$.

Further, we have

$$E_{-\theta_0}e^{2\theta_0 R_\infty} = e^{2\theta_0\rho}(1+o(\theta_0^2)),$$

by using the facts

$$E_{-\theta_0}R_\infty = \rho - \frac{\theta_0}{4}+o(\theta_0) \quad \text{and} \quad E_0 R_\infty^2 = \rho^2 + \frac{1}{4}.$$

Thus, locally,

$$\frac{\frac{\partial}{\partial\theta_0}\left(\theta_0 E_{-\theta_0} e^{2\theta_0 R_\infty} P_{-\theta_0}(\tau_- = \infty)\right)}{E_{-\theta_0} e^{2\theta_0 R_\infty} P_{-\theta_0}(\tau_- = \infty)} = \frac{\frac{\partial}{\partial\theta_0}(\theta_0^2 e^{3\theta_0 \rho})}{\theta_0 e^{3\theta_0 \rho}}(1 + o(1))$$

$$= 2 + 3\theta_0 \rho + o(\theta_0).$$

The Taylor series expansion gives

$$E_{-\theta_0} R_\infty e^{2\theta_0 R_\infty} = \rho + \theta_0(2\rho^2 + \frac{1}{4}) + o(\theta_0).$$

Finally, we have the next result.

Corollary 5.1: *Locally,*

$$E_{\theta_0}\left(\frac{S_{N_0}}{N_0} + \theta_0 | S_{N_0} > d\right) = \frac{1}{d}\left(2 + \theta_0 \rho - \frac{\theta_0^2}{2} + o(\theta_0^2)\right)(1 + o(1)). \qquad (5.6)$$

5.2 Corrected Normal Approximation

Given $S_{N_0} > d$, Bertoin and Doney (1994) showed that as $d \to \infty$, $\{S_n\}$ behaves like a Markov chain with transition probability

$$P^*(x, dy) = e^{-2\theta_0(y-x)}\phi(y - x + \theta_0)dy.$$

Thus, the normality of the sample mean S_{N_0}/N_0 can be established with the Markov property.

In the following, we concentrate on obtaining a second-order corrected normal approximation for

$$Z_{N_0} = \sqrt{N_0}\left(\frac{S_{N_0}}{N_0} + \theta_0\right)$$

by finding its bias and variance. The idea has been studied in Chapter 4. The technique used here is an extension of the ones used in Chapter 4.

5.2.1 Bias of Z_{N_0}

Following the similar technique in the previous section, we first write

$$2E_{\theta_0}[Z_{N_0}|S_{N_0} > d]$$

$$= 2E_{\theta_0}\left[\frac{S_{N_0} - d}{\sqrt{N_0}}|S_{N_0} > d\right] + 2E_{\theta_0}\left[\sqrt{N_0}\left(\frac{d}{N_0} + \theta_0\right)|S_{N_0} > d\right]. \qquad (5.7)$$

As the second term of (5.7) is at the same order as $E_{\theta_0}[Z_{N_0}|S_{N_0} > d]$, we need another equation to eliminate it. By using the normal density property, we have

$$E_{\theta_0}[Z_{N_0}|S_{N_0} > d] = E_{\theta_0}\left[\frac{1}{\sqrt{N_0}}(S_{N_0} - N_0\theta_0)|S_{N_0} > d\right] + 2\theta_0 E_{\theta_0}[\sqrt{N_0}|S_{N_0} > d]$$

$$= \frac{\frac{\partial}{\partial\theta_0}E_{\theta_0}\left[\frac{1}{\sqrt{N_0}}; S_{N_0} > d\right]}{P_{\theta_0}(S_{N_0} > d)} + 2\theta_0 E_{\theta_0}[\sqrt{N_0}|S_{N_0} > d]$$

$$= \frac{\frac{\partial}{\partial\theta_0}E_{-\theta_0}\left[e^{2\theta_0 d}\frac{1}{\sqrt{N_0}}e^{2\theta_0(S_{N_0}-d)}; S_{N_0} > d\right]}{e^{2\theta_0 d}E_{-\theta_0}\left[e^{2\theta_0(S_{N_0}-d)}; S_{N_0} > d\right]} + 2\theta_0 E_{\theta_0}[\sqrt{N_0}|S_{N_0} > d]$$

$$= 2dE_{\theta_0}\left[\frac{1}{\sqrt{N_0}}|S_{N_0} > d\right] + \frac{\frac{\partial}{\partial\theta_0}E_{-\theta_0}\left[\frac{1}{\sqrt{N_0}}e^{2\theta_0(S_{N_0}-d)}; S_{N_0} > d\right]}{E_{-\theta_0}\left[e^{2\theta_0(S_{N_0}-d)}; S_{N_0} > d\right]}$$

$$+ 2\theta_0 E_{\theta_0}[\sqrt{N_0}|S_{N_0} > d]$$

$$= 2\theta_0 E_{\theta_0}\left[\sqrt{N_0}\left(\theta_0 + \frac{d}{N_0}\right)|S_{N_0} > d\right]$$

$$+ \frac{\frac{\partial}{\partial\theta_0}E_{-\theta_0}\left[\frac{1}{\sqrt{N_0}}e^{2\theta_0(S_{N_0}-d)}; S_{N_0} > d\right]}{E_{-\theta_0}\left[e^{2\theta_0(S_{N_0}-d)}; S_{N_0} > d\right]}. \tag{5.8}$$

By subtracting (5.8) from (5.7), we get

$$E_{\theta_0}(Z_{N_0}|S_{N_0} > d) = 2E_{\theta_0}\left[\frac{S_{N_0} - d}{\sqrt{N_0}}|S_{N_0} > d\right]$$

$$- \frac{\frac{\partial}{\partial\theta_0}E_{-\theta_0}\left[\frac{1}{\sqrt{N_0}}e^{2\theta_0(S_{N_0}-d)}; S_{N_0} > d\right]}{E_{-\theta_0}\left[e^{2\theta_0(S_{N_0}-d)}; S_{N_0} > d\right]}.$$

Thus, we have the following asymptotic result:

Theorem 5.2: *As* $d \to \infty$,

$$E_{\theta_0}(Z_{N_0}|S_{N_0} > d) = \frac{a(\theta_0)}{\sqrt{d}}(1 + o(1)),$$

where

$$a(\theta_0) = 2\sqrt{-\theta_0}\frac{E_{-\theta_0}\left(R_\infty e^{2\theta_0 R_\infty}\right)}{E_{-\theta_0}e^{2\theta_0 R_\infty}} - \frac{\frac{\partial}{\partial\theta_0}\left(\sqrt{-\theta_0}E_{-\theta_0}e^{2\theta_0 R_\infty}P_{-\theta_0}(\tau_- = \infty)\right)}{E_{-\theta_0}e^{2\theta_0 R_\infty}P_{-\theta_0}(\tau_- = \infty)}.$$

In the local case as $\theta_0 \to 0$, a similar algebraic computation with the approximations given for Corollary 5.1 shows the following simple approximation.

Corollary 5.2: *As* $d \to \infty$ *and then* $\theta_0 \to 0$,

$$E_{\theta_0}(Z_{N_0}|S_{N_0} > d) = \frac{1}{\sqrt{-d\theta_0}}\left(\frac{3}{2} + \theta_0\rho - \frac{\theta_0^2}{2} + o(\theta_0^2)\right)(1 + o(1)).$$

5.2.2   Variance of Z_{N_0}

Here we evaluate

$$Var_{\theta_0}(Z_{N_0}|S_{N_0} > d) = E_{\theta_0}[Z_{N_0}^2|S_{N_0} > d] - \frac{a^2(\theta_0)}{d}.$$

The derivation needs more careful techniques. First, we write

$$4E_{\theta_0}[Z_{N_0}^2|S_{N_0} > d] = 4E_{\theta_0}\left[N_0\left(\frac{S_{N_0}}{N_0} - \frac{d}{N_0} + \frac{d}{N_0} + \theta_0\right)^2|S_{N_0} > d\right]$$

$$= 4E_{\theta_0}\left[\frac{1}{N_0}(S_{N_0} - d)^2|S_{N_0} > d\right] + 8E_{\theta_0}\left[(S_{N_0} - d)\left(\frac{d}{N_0} + \theta_0\right)|S_{N_0} > d\right]$$

$$+ 4\theta_0^2 E_{\theta_0}\left[\frac{1}{N_0}\left(N_0 + \frac{d}{\theta_0}\right)^2|S_{N_0} > d\right]. \tag{5.9}$$

As the third term in (5.9) is at the same order as $E_{\theta_0}[Z_{N_0}^2|S_{N_0} > d]$, we need to find another equation to eliminate it by extending the techniques in the previous subsection.

By using the normal density property and Wald's likelihood ratio identity, we write

$$E_{\theta_0}[Z_{N_0}^2|S_{N_0} > d] = E_{\theta_0}\left[N_0\left(\frac{S_{N_0}}{N_0} - \theta_0 + 2\theta_0\right)^2|S_{N_0} > d\right]$$

$$= E_{\theta_0}\left[N_0\left(\frac{S_{N_0}}{N_0} - \theta_0\right)^2|S_{N_0} > d\right]$$

$$+ 4\theta_0 E_{\theta_0}\left[N_0\left(\frac{S_{N_0}}{N_0} - \theta_0\right)|S_{N_0} > d\right] + 4\theta_0^2 E_{\theta_0}[N_0|S_{N_0} > d]$$

$$= E_{\theta_0}\left[N_0\left(\frac{S_{N_0}}{N_0} - \theta_0\right)^2|S_{N_0} > d\right]$$

$$+ 4\theta_0 \frac{\frac{\partial}{\partial\theta_0}P_{\theta_0}(S_N > d)}{P_{\theta_0}(S_{N_0} > d)} + 4\theta_0^2 E_{\theta_0}[N_0|S_{N_0} > d]$$

$$= E_{\theta_0}\left[N_0\left(\frac{S_{N_0}}{N_0} - \theta_0\right)^2|S_{N_0} > d\right]$$

$$+ 4\theta_0 \frac{\frac{\partial}{\partial\theta_0}E_{-\theta_0}\left[e^{2\theta_0 d}e^{2\theta_0(S_{N_0} - d)}; S_{N_0} > d\right]}{E_{-\theta_0}\left[e^{2\theta_0 d}e^{2\theta_0(S_{N_0} - d)}; S_{N_0} > d\right]} + 4\theta_0^2 E_{\theta_0}[N_0|S_{N_0} > d]$$

$$= E_{\theta_0}\left[N_0\left(\frac{S_{N_0}}{N_0} - \theta_0\right)^2|S_{N_0} > d\right] + 8\theta_0 d + 4\theta_0^2 E_{\theta_0}[N_0|S_{N_0} > d]$$

$$+ 4\theta_0 \frac{\partial}{\partial\theta_0}\ln\left[E_{-\theta_0}e^{2\theta_0 R_\infty}P_{-\theta_0}(\tau_- = \infty)\right](1 + o(1)).$$

For the first term, we use the normal density property again and get

$$E_{\theta_0}\left[N_0\left(\frac{S_{N_0}}{N_0}-\theta_0\right)^2|S_{N_0}>d\right]=1+\frac{1}{P_{\theta_0}(S_{N_0}>d)}\frac{\partial^2}{\partial^2\theta_0}E_{\theta_0}\left[\frac{1}{N_0};S_{N_0}>d\right]$$

$$=1+\frac{1}{P_{\theta_0}(S_{N_0}>d)}\frac{\partial^2}{\partial^2\theta_0}\left[e^{2\theta_0d}E_{-\theta_0}\left(\frac{1}{N_0}e^{2\theta_0(S_{N_0}-d)};S_{N_0}>d\right)\right]$$

$$=1+\frac{1}{P_{\theta_0}(S_{N_0}>d)}\{4d^2e^{2\theta_0d}E_{-\theta_0}\left[\frac{1}{N_0}e^{2\theta_0(S_{N_0}-d)};S_{N_0}>d\right]$$

$$+4de^{2\theta_0d}\frac{\partial}{\partial\theta_0}E_{-\theta_0}\left[\frac{1}{N_0}e^{2\theta_0(S_{N_0}-d)};S_{N_0}>d\right]$$

$$+e^{2\theta_0d}\frac{\partial^2}{\partial^2\theta_0}E_{-\theta_0}\left[\frac{1}{N_0}e^{2\theta_0(S_{N_0}-d)};S_{N_0}>d\right]\}$$

$$=1+4d^2E_{\theta_0}\left[\frac{1}{N_0}|S_{N_0}>d\right]+4d\frac{\frac{\partial}{\partial\theta_0}E_{-\theta_0}\left[\left(\frac{1}{N_0}+\frac{\theta_0}{d}\right)e^{2\theta_0(S_{N_0}-d)};S_{N_0}>d\right]}{E_{-\theta_0}\left[e^{2\theta_0(S_{N_0}-d)};S_{N_0}>d\right]}$$

$$-4\frac{\frac{\partial}{\partial\theta_0}\left(\theta_0E_{-\theta_0}\left[e^{2\theta_0(S_{N_0}-d)};S_{N_0}>d\right]\right)}{E_{-\theta_0}\left[e^{2\theta_0(S_{N_0}-d)};S_{N_0}>d\right]}$$

$$+\frac{\frac{\partial^2}{\partial^2\theta_0}E_{-\theta_0}\left[\frac{1}{N_0}e^{2\theta_0(S_{N_0}-d)};S_{N_0}>d\right]}{E_{-\theta_0}\left[e^{2\theta_0(S_{N_0}-d)};S_{N_0}>d\right]}$$

$$=1+4d^2E_{\theta_0}\left[\frac{1}{N_0}|S_{N_0}>d\right]+4d\frac{\frac{\partial}{\partial\theta_0}E_{-\theta_0}\left[\left(\frac{1}{N_0}+\frac{\theta_0}{d}\right)e^{2\theta_0(S_{N_0}-d)};S_{N_0}>d\right]}{E_{-\theta_0}\left[e^{2\theta_0(S_{N_0}-d)};S_{N_0}>d\right]}$$

$$-4-4\theta_0\frac{\partial}{\partial\theta_0}\ln\left[E_{-\theta_0}e^{2\theta_0R_\infty}P_{-\theta_0}(\tau_-=\infty)\right]$$

$$-\frac{1}{d}\frac{\frac{\partial^2}{\partial^2\theta_0}\left[\theta_0E_{-\theta_0}e^{2\theta_0R_\infty}P_{-\theta_0}(\tau_-=\infty)\right]}{E_{-\theta_0}e^{2\theta_0R_\infty}P_{-\theta_0}(\tau_-=\infty)}(1+o(1)).$$

Thus,

$$E_{\theta_0}[Z_{N_0}^2|S_{N_0}>d]=-3+4d^2E_{\theta_0}\left[\frac{1}{N_0}\left(1+\frac{\theta_0N_0}{d}\right)^2|S_{N_0}>d\right]$$

$$+4d\frac{\frac{\partial}{\partial\theta_0}E_{-\theta_0}\left[\left(\frac{1}{N_0}+\frac{\theta_0}{d}\right)e^{2\theta_0(S_{N_0}-d)};S_{N_0}>d\right]}{E_{-\theta_0}\left[e^{2\theta_0(S_{N_0}-d)};S_{N_0}>d\right]}$$

$$-\frac{1}{d}\frac{\frac{\partial^2}{\partial^2\theta_0}\left[\theta_0E_{-\theta_0}e^{2\theta_0R_\infty}P_{-\theta_0}(\tau_-=\infty)\right]}{E_{-\theta_0}e^{2\theta_0R_\infty}P_{-\theta_0}(\tau_-=\infty)}. \qquad (5.10)$$

Subtracting (5.10) from (5.9), we get

$$E_{\theta_0}[Z_{N_0}^2|S_{N_0}>d]=1+\frac{4}{3}E_{\theta_0}\left[\frac{1}{N_0}(S_{N_0}-d)^2|S_{N_0}>d\right]$$

$$+\frac{8}{3}E_{\theta_0}\left[(S_{N_0}-d)\left(\frac{d}{N_0}+\theta_0\right)|S_{N_0}>d\right]$$

$$-\frac{4d}{3}\frac{\frac{\partial}{\partial\theta_0}E_{-\theta_0}\left[\left(\frac{1}{N_0}+\frac{\theta_0}{d}\right)e^{2\theta_0(S_{N_0}-d)}; S_{N_0}>d\right]}{E_{-\theta_0}\left[e^{2\theta_0(S_{N_0}-d)}; S_{N_0}>d\right]}$$

$$+\frac{1}{3d}\frac{\frac{\partial^2}{\partial^2\theta_0}\left[\theta_0 E_{-\theta_0}e^{2\theta_0 R_\infty}P_{-\theta_0}(\tau_-=\infty)\right]}{E_{-\theta_0}e^{2\theta_0 R_\infty}P_{-\theta_0}(\tau_-=\infty)}(1+o(1)).$$

The second term can be evaluated as

$$E_{\theta_0}\left[\frac{1}{N_0}(S_{N_0}-d)^2|S_{N_0}>d\right]=-\frac{\theta_0}{d}\frac{E_{-\theta_0}\left(R_\infty^2 e^{2\theta_0 R_\infty}\right)}{E_{-\theta_0}e^{2\theta_0 R_\infty}}(1+o(1)).$$

By using the byproduct given in (5.5), the fourth term can be obtained as

$$\frac{\partial}{\partial\theta_0}E_{-\theta_0}\left[\left(\frac{1}{N_0}+\frac{\theta_0}{d}\right)e^{2\theta_0(S_{N_0}-d)}; S_{N_0}>d\right]$$

$$=\frac{1}{d^2}\{\frac{\partial^2}{\partial^2\theta_0}(\theta_0 E_{-\theta_0}e^{2\theta_0 R_\infty}P_{-\theta_0}(\tau_-=\infty))$$

$$-\frac{\partial}{\partial\theta_0}(\theta_0 E_{-\theta_0}\left(R_\infty e^{2\theta_0 R_\infty}\right)P_{-\theta_0}(\tau_-=\infty))\}(1+o(1)).$$

The third term can be evaluated as follows:

$$E_{\theta_0}\left[(S_{N_0}-d)\left(\frac{d}{N_0}+\theta_0\right)|S_{N_0}>d\right]$$

$$=-\frac{\theta_0^2}{d}E_{\theta_0}\left[(S_{N_0}-d)\left(N_0+\frac{d}{\theta_0}\right)|S_{N_0}>d\right](1+o(1))$$

$$=-\frac{\theta_0^2}{d}\frac{E_{\theta_0}\left[(S_{N_0}-d)\left(N_0+\frac{d}{\theta_0}\right); S_{N_0}>d\right]}{P_{\theta_0}(S_{N_0}>d)}(1+o(1))$$

$$=-\frac{\theta_0^2}{d}\frac{E_{-\theta_0}\left[(S_{N_0}-d)\left(N_0+\frac{d}{\theta_0}\right)e^{2\theta_0(S_{N_0}-d)}; S_{N_0}>d\right]}{P_{\theta_0}(S_{N_0}>d)}(1+o(1))$$

$$=-\frac{\theta_0^2}{d}\frac{E_{-\theta_0}\left(R_d\left(\tau_d+\frac{d}{\theta_0}\right)e^{2\theta_0 R_d}\right)-E_{-\theta_0}\left[R_d\left(\tau_d+\frac{d}{\theta_0}\right)e^{2\theta_0 R_d}; S_{N_0}\le 0\right]}{P_{\theta_0}(S_{N_0}>d)}$$

$$\times(1+o(1))$$

$$=\frac{\theta_0^2}{d}\{\frac{\lim E_{-\theta_0}\left(R_d\left(\tau_d+\frac{d}{\theta_0}\right)e^{2\theta_0 R_d}\right)P_{-\theta_0}(\tau_-=\infty)}{E_{-\theta_0}e^{2\theta_0 R_\infty}P_{-\theta_0}(\tau_-=\infty)}$$

$$-\frac{E_{-\theta_0}\left(R_\infty e^{2\theta_0 R_\infty}\right)E_{-\theta_0}\left[N_0+\frac{S_{N_0}}{\theta_0}; S_{N_0}\le 0\right]}{E_{-\theta_0}e^{2\theta_0 R_\infty}P_{-\theta_0}(\tau_-=\infty)}\}$$

$$\times(1+o(1))$$

$$=-\frac{\theta_0^2}{d}\{\frac{\lim E_{-\theta_0}\left(R_d\left(\tau_d+\frac{d}{\theta_0}\right)e^{2\theta_0 R_d}\right)P_{-\theta_0}(\tau_-=\infty)}{E_{-\theta_0}e^{2\theta_0 R_\infty}P_{-\theta_0}(\tau_-=\infty)}$$

$$-\frac{E_{-\theta_0}\left(R_\infty e^{2\theta_0 R_\infty}\right)E_{-\theta_0}\left[\tau_-+\frac{S_{\tau_-}}{\theta_0}; \tau_-<\infty\right]}{E_{-\theta_0}e^{2\theta_0 R_\infty}P_{-\theta_0}(\tau_-=\infty)}\}(1+o(1)),$$

where for the second term in the second equation from the last, we use the fact that given $S_{N_0} = x \leq 0$ and $N_0 = k$, τ_d is equivalent to $k + \tau_{d-x}$, R_d changes to R_{d-x}, and d is replaced by $d - x + x$. The limit is taken as $d \to \infty$. Further, by using Wald's likelihood ratio identity by changing $P_{-\theta_0}(.)$ to $P_0(.)$, we note that

$$\frac{\partial}{\partial \theta_0} E_{-\theta_0} \left(R_d e^{2\theta_0 R_d} \right) = \frac{\partial}{\partial \theta_0} E_0 \left[R_d e^{2\theta_0 R_d} e^{-\theta_0 S_{\tau_d} - \tau_d \theta_0^2 / 2} \right]$$

$$= 2 E_{-\theta_0} \left(R_d^2 e^{2\theta_0 R_d} \right) - E_{-\theta_0} \left[R_d (S_{\tau_d} + \tau_d \theta_0) e^{2\theta_0 R_d} \right]$$

$$= E_{-\theta_0} \left(R_d^2 e^{2\theta_0 R_d} \right) - \theta_0 E_{-\theta_0} \left[R_d \left(\tau_d + \frac{d}{\theta_0} \right) e^{2\theta_0 R_d} \right],$$

and a similar idea has been used in Chang (1992).

Thus,

$$\lim_{d \to \infty} E_{-\theta_0} \left[R_d \left(\tau_d + \frac{d}{\theta_0} \right) e^{2\theta_0 R_d} \right]$$

$$= -\frac{1}{\theta_0} \left[E_{-\theta_0} \left(R_\infty^2 e^{2\theta_0 R_\infty} \right) - \frac{\partial}{\partial \theta_0} E_{-\theta_0} \left(R_\infty e^{2\theta_0 R_\infty} \right) \right].$$

Combining the above asymptotic results, we finally get

Theorem 5.3: *As $d \to \infty$,*

$$E_{\theta_0} [Z_{N_0}^2 | S_{N_0} > d] = 1 + \frac{b(\theta_0)}{d} (1 + o(1)),$$

where

$$b(\theta_0) = -\frac{\frac{\partial^2}{\partial^2 \theta_0} \left[\theta_0 E_{-\theta_0} e^{2\theta_0 R_\infty} P_{-\theta_0} (\tau_- = \infty) \right]}{E_{-\theta_0} e^{2\theta_0 R_\infty} P_{-\theta_0} (\tau_- = \infty)}$$

$$+ \frac{4}{3} \frac{\frac{\partial}{\partial \theta_0} \left(\theta_0 E_{-\theta_0} \left(R_\infty e^{2\theta_0 R_\infty} \right) P_{-\theta_0} (\tau_- = \infty) \right)}{E_{-\theta_0} e^{2\theta_0 R_\infty} P_{-\theta_0} (\tau_- = \infty)}$$

$$+ \frac{8 \theta_0^2}{3} \frac{E_{-\theta_0} \left(R_\infty e^{2\theta_0 R_\infty} \right) E_{-\theta_0} \left[\tau_- + \frac{S_{\tau_-}}{\theta_0} ; \tau_- < \infty \right]}{E_{-\theta_0} e^{2\theta_0 R_\infty} P_{-\theta_0} (\tau_- = \infty)}$$

$$+ \frac{4 \theta_0}{3} \frac{E_{-\theta_0} \left(R_\infty^2 e^{2\theta_0 R_\infty} \right)}{E_{-\theta_0} e^{2\theta_0 R_\infty}} - \frac{8 \theta_0}{3} \frac{\frac{\partial}{\partial \theta_0} E_{-\theta_0} R_\infty e^{2\theta_0 R_\infty}}{E_{-\theta_0} e^{2\theta_0 R_\infty}}.$$

Locally, we further assume $\theta_0 \to 0$. We have

$$-\frac{\frac{\partial^2}{\partial^2 \theta_0} \left[\theta_0 E_{-\theta_0} e^{2\theta_0 R_\infty} P_{-\theta_0} (\tau_- = \infty) \right]}{E_{-\theta_0} e^{2\theta_0 R_\infty} P_{-\theta_0} (\tau_- = \infty)} = -\frac{\frac{\partial}{\partial \theta_0} \left(\theta_0^2 e^{3\theta_0 \rho} \right)}{\theta_0 e^{3\theta_0 \rho}} (1 + o(1))$$

$$= -\frac{1}{\theta_0} (2 + 12 \rho \theta_0 + 9 \rho^2 \theta_0^2)(1 + o(1)),$$

$$\frac{\frac{\partial}{\partial\theta_0}\left(\theta_0 E_{-\theta_0}\left(R_\infty e^{2\theta_0 R_\infty}\right)P_{-\theta_0}(\tau_-=\infty)\right)}{E_{-\theta_0}e^{2\theta_0 R_\infty}P_{-\theta_0}(\tau_-=\infty)}$$

$$= \frac{\frac{\partial}{\partial\theta_0}\left(\theta_0^2 e^{\theta_0\rho}\left(\rho+\left(\frac14+2\rho^2\right)\theta_0\right)\right)}{\theta_0 e^{3\theta_0\rho}}(1+o(1))$$

$$= 2\rho-\theta_0\left(\rho^2-\frac14\right)+o(\theta_0).$$

Furthermore, we have

$$E_{-\theta_0}\left[\tau_-+\frac{S_{\tau_-}}{\theta_0};\tau_-<\infty\right] = E_{\theta_0}\left[(\tau_-+\frac{S_{\tau_-}}{\theta_0})e^{-2\theta_0 S_{\tau_-}}\right]$$

$$= E_{\theta_0}\left(\tau_-+\frac{S_{\tau_-}}{\theta_0}\right)-2\theta_0 E_{\theta_0}\left(\tau_- S_{\tau_-}+\frac{S_{\tau_-}^2}{\theta_0}\right)$$

$$= -\frac{2}{\theta_0}\left(E_0 S_{\tau_+}-\frac{\theta_0}{2}E_0 S_{\tau_+}^2+o(\theta_0)\right)$$

$$\quad - 3E_0 S_{\tau_+}^2+o(1)$$

$$= -\frac{2}{\theta_0}E_0 S_{\tau_+}(1+2\rho\theta_0+o(\theta_0)),$$

where we used Lemma 10.27 of Siegmund (1985).

Thus,

$$\frac{8\theta_0^2}{3}\frac{E_{-\theta_0}\left(R_\infty e^{2\theta_0 R_\infty}\right)E_{-\theta_0}\left[\tau_-+\frac{S_{\tau_-}}{\theta_0};\tau_-<\infty\right]}{E_{-\theta_0}e^{2\theta_0 R_\infty}P_{-\theta_0}(\tau_-=\infty)}$$

$$= \frac{8\theta_0^2}{3}\frac{(\rho+\theta_0(2\rho^2+1/2))(-2E_0 S_{\tau_+}(1+2\rho\theta_0)/\theta_0)}{-\theta_0 e^{3\theta_0\rho}/E_0 S_{\tau_+}}(1+o(1))$$

$$= \frac{8}{3}\left(\rho+\theta_0\left(\rho^2+\frac14\right)+o(\theta_0)\right).$$

Next, we note that

$$\frac{4\theta_0}{3}\frac{E_{-\theta_0}\left(R_\infty^2 e^{2\theta_0 R_\infty}\right)}{E_{-\theta_0}e^{2\theta_0 R_\infty}} = \frac{4\theta_0}{3}\left(\rho^2+\frac14\right)+o(\theta_0),$$

and

$$-\frac{8\theta_0}{3}\frac{\frac{\partial}{\partial\theta_0}E_{-\theta_0}\left(R_\infty e^{2\theta_0 R_\infty}\right)}{E_{-\theta_0}e^{2\theta_0 R_\infty}} = -\frac{8\theta_0}{3}\frac{\partial}{\partial\theta_0}(\rho+\theta_0(2\rho^2+1/4))(1+o(1))$$

$$= -\frac{8\theta_0}{3}\left(2\rho^2+\frac14\right)(1+o(1)).$$

Thus, we have the following local approximation.

Table 5.1: Simulation Comparison with $d = 5$

θ_0	Runs	$P(S_{N_0} > d)$	$E^*[\hat{\theta} + \theta_0]$	$E^*[Z_{N_0}]$	$E^*[Z_{N_0}^2]$
-0.5	150,000	284/150,000	0.300	0.619	1.089
		(0.002)	(0.317)	(0.685)	(1.352)
-0.25	50,000	965/50,000	0.411	1.040	1.610
		(0.019)	(0.365)	(1.183)	(1.987)
-0.1	50,000	3026/50,000	0.503	1.423	2.390
		(0.044)	(0.387)	(2.032)	(4.288)

Corollary 5.3: *As $\theta_0 \to 0$, the second moment of Z_{N_0} is*

$$E_{\theta_0}[Z_{N_0}^2|S_{N_0} > d] = 1 - \frac{1}{d\theta_0}\left(2 + \frac{20}{3}\rho\theta_0 + \theta_0^2\left(\frac{35}{3}\rho^2 - \frac{2}{3}\right) + o(\theta_0^2)\right)(1 + o(1)),$$

and the conditional variance is

$$\text{Var}_{\theta_0}(Z_{N_0}|S_{N_0} > d) = 1 + \frac{1}{d\theta_0}\left(\frac{1}{4} - \frac{11}{3}\rho\theta_0 - \theta_0^2\left(\frac{35}{3}\rho^2 - \frac{1}{6}\right) + o(\theta_0^2)\right)(1 + o(1)).$$

Now, a corrected normal approximation for Z_{N_0} can be formed as

$$Z^* = (Z_{N_0} - a(\theta_0)/\sqrt{d})/(1 + (b(\theta_0) - a^2(\theta_0))/d)^{1/2}.$$

5.3   Numerical Comparison and Discussion

In this section, we run a simple simulation study. We let $d = 5$ and $\theta_0 = -0.5, -0.25$ and -0.1. We simulate the normal random walk $\{S_n\}$ until the random stopping time N. Only those with $S_{N_0} > d$ are recorded and used to estimate the mean of S_{N_0}/N_0 and the bias and second moment of Z_{N_0}.

The following table gives the simulation results and also reports the number of runs and the probability $P_{\theta_0}(S_{N_0} > d)$. The approximated value for $P_{\theta_0}(S_{N_0} > d)$ can be obtained from Section 5.2 as

$$P_{\theta_0}(S_{N_0} > d) = -\sqrt{2}\theta_0 e^{2\theta_0 d + 3\theta_0 \rho}(1 + o(1)).$$

The approximated values for $E_{\theta_0}[S_{N_0}/N_0 + \theta_0|S_{N_0} > d]$ and the first two moments for Z_{N_0} are calculated from Corollaries 5.1, 5.2, and 5.3 and given in the parentheses. We use $E^*[.]$ to denote $E[.|S_{N_0} > d]$.

From the table, we see that the local approximations are quite satisfactory for $\theta_0 = -0.5$ and -0.25. Note that we first let $d \to \infty$ and then $\theta_0 \to 0$. Thus, the approximation when $\theta_0 = -0.1$ does not perform very well as only the first-order asymptotic result is given and the approximations for the bias and variance of Z_{N_0} are given at the order of $1/(-d\theta_0)$. In the simulation study, $d =$

5 which is not too large in order to save simulation time. A moderate deviation analysis by assuming $d\theta_0 \to$ constant will help to get more accurate results. The numerical results show that the bias correction is definitely necessary.

Finally, there are a few points related to the application of the results. For classifying the true signal and false signal point of view, the results can be used in the following way. Suppose $-\theta_0$ is the minimum post-change mean we want to detect and the true mean is $\theta > -\theta_0$. Then the marginal distribution for the post-change mean without knowing the status of a signal is a mixture of the two conditional distributions from the false and true signal, respectively.

For the design of the alarming limit d, two factors affect the selection: change magnitude and changed segment length. If we take the minimum magnitude (reference) as $-\theta_0$, then d should be selected such that the delay detection time under the change magnitude $-\theta_0$ is shorter than the segment length in order to detect the changed segments.

6

Inference with Change in Variance

6.1 Introduction

We consider the following change-point problem in a sequence of independent normal random variables $\{X_i\}$ for $i = 1, 2, \dots$. Suppose for $k \leq \nu$, X_k follows $N(\theta_0, 1)$ and for $k > \nu$, X_k follows $N(\theta, \sigma^2)$, where ν is the change-point and $\theta_0 < 0 < \theta$. Our initial goal is to detect changes in the mean by using the CUSUM procedure with reference value $-\theta_0$. However, as the post-change mean is rarely known and also it is possible that the variance is subject to change, we estimate the post-change mean and variance as

$$\hat{\theta} = \frac{1}{N - \hat{\nu}} T_N = \frac{1}{T - \hat{\nu}} \sum_{i=\hat{\nu}+1}^{N} X_i$$

and

$$\hat{\sigma}^2 = \frac{1}{N - \hat{\nu}} \sum_{i=\hat{\nu}+1}^{N} (X_i - \hat{\theta})^2,$$

which are the ordinary maximum likelihood estimators when $\theta = \theta_0$.

In the case $\sigma^2 = 1$ without a change, conditional on $N > \nu$, Chapters 2 and 4 consider the biases of $\hat{\nu}$ and $\hat{\theta}$ and a confidence interval for θ is constructed based on the corrected normal pivot. In this chapter, we generalize the results to the case when the variance is also subject to change. The study serves two purposes. First, we want to deal with more practical situations. Second, we want to study the robustness of the estimator for the post-change mean when the variance is unknown. Although the method used here follows the lines of Chapters 2 and 4, the derivation raises new technical difficulties. In Section 6.2, we consider the (absolute) bias of $\hat{\nu}$ conditional on $N > \nu$ by assuming $\nu, d \to \infty$. Approximation in the local case for small values of θ_0 and θ is obtained. The results generalize the ones given in Chapter 2. In Section 6.3, we consider the biases of $\hat{\theta}$ and $\hat{\sigma}^2$. The techniques used here are extensions of Chapter 4 and Coad and Woodroofe (1998). These results and techniques are then used to construct a confidence interval for the post-change mean in Section 6.4 by using a corrected normal pivot. Numerical evaluations and a real example are presented in Section 6.5. Necessary technical details are given in the appendix.

6.2   Change-Point Estimation

For the convenience of presentation, we first introduce some standard notations.

Denote $P_{(\theta,\sigma)}(.)$ as the measure under the parameter (θ, σ^2) and $P_\theta(.)$ the measure when $\sigma = 1$. Denote

$$m = \inf_{n \geq 0} S_n \quad \text{and} \quad \gamma_m = \arg\inf_{n \geq 0} S_n.$$

For an independent copy $\{S'_n\}$ of $\{S_n\}$ under $P_{\theta_0}(.)$, we define

$$M = \sup_{n \geq 0} S'_n; \quad \text{and} \quad \eta_M = \mathrm{argsup}_{n \geq 0} S'_n,$$

for $S'_0 = 0$.

When both parameters $(\theta_0, 1)$ and (θ, σ) are involved, we simply denote by $P(.)$ the joint measure. Also, we denote $P^\nu(.)$ as the induced measure when a change occurs at ν.

For the change-point estimation, the most important properties are the bias and absolute bias of $\hat{\nu}$ as ν is a time index. Wu (1999) considers the bias of $\hat{\nu}$ in the fixed sample size case, and Chapter 2 further extends to the sequential sampling plan case after a CUSUM test in a one-parameter exponential family.

As $\nu, d \to \infty$, conditional on $N > \nu$, the following asymptotic result can be obtained as in Chapter 2 by considering the case $\hat{\nu} > \nu$ and $\hat{\nu} < \nu$ separately.

Theorem 6.1: *As $\nu, d \to \infty$,*

$$E^\nu[\hat{\nu} - \nu | N > \nu] \quad \to \quad E[\tau_{-M}; \tau_{-M} < \infty]$$

$$+ P(\tau_{-M} < \infty) \frac{E_{(\theta,\sigma)}[\tau_-; \tau_- < \infty]}{P_{(\theta,\sigma)}(\tau_- = \infty)} - E[\eta_M; \tau_{-M} = \infty]$$

and

$$E^{\nu}[\|\hat{\nu} - \nu\| | N > \nu] \quad \rightarrow \quad E[\tau_{-M}; \tau_{-M} < \infty]$$
$$+ P(\tau_{-M} < \infty)\frac{E_{(\theta,\sigma)}[\tau_-; \tau_- < \infty]}{P_{(\theta,\sigma)}(\tau_- = \infty)} + E[\eta_M; \tau_{-M} = \infty].$$

In the following, we obtain a second-order expansion for the bias and the absolute bias by further assuming both θ_0 and θ approach zero at the same order for a numerical evaluation.

The technique used here is the uniform strong renewal theorem as discussed in Chapter 1.3. Most results are direct generalizations of Wu (1999) and Chapter 2. The main steps are presented in a sequence of lemmas.

The first lemma gives an approximation for the probability that a random walk will never go below zero and the length of the descent ladder time with positive drift θ and diffusion parameter σ.

Lemma 6.1: *As $\theta \to 0$,*

$$P_{(\theta,\sigma)}(\tau_- = \infty) = \sqrt{2}\frac{\theta}{\sigma}e^{-\theta\rho/\sigma}(1 + O(\theta^3)),$$
$$E_{(\theta,\sigma)}[\tau_-; .\tau_- < \infty] = \frac{\sigma}{\sqrt{2\theta}}e^{-\theta\rho/\sigma-(\theta/\sigma)^2/2}(1 + O(\theta^3)),$$

where $\rho = E_0 R_\infty \approx 0.583$.

The proof is obtained by noting that

$$P_{(\theta,\sigma)}(\tau_- = \infty) = P_{\theta/\sigma}(\tau_- = \infty)$$

and

$$E_{(\theta,\sigma)}[\tau_-; \tau_- < \infty] = E_{\theta/\sigma}[\tau_-; \tau_- < \infty].$$

Since γ_m can be seen as a geometric summation of i.i.d. variables distributed as $[\tau_-; \tau_- < \infty]$ with terminating probability $P_{(\theta,\sigma)}(\tau_- = \infty)$, we have

$$E_{(\theta,\sigma)}(\gamma_m) = \frac{E_{(\theta,\sigma)}[\tau_-; \tau_- < \infty]}{P_{(\theta,\sigma)}(\tau_- = \infty)}$$
$$= \frac{1}{2(\theta/\sigma)^2} - \frac{1}{4} + o(1),$$

as in Lemma 5 of Wu (1999).

The second lemma gives an approximation to the probability that the CUSUM process will go back to zero after the change has occurred. The technique generalizes Lemma 6 of Wu (1999). For the sake of easy reading, we present a brief proof in the appendix.

Lemma 6.2: *As $\theta_0, \theta \to 0$ at the same order,*

$$P(\tau_{-M} < \infty)$$

$$= -\frac{\theta_0}{\theta/\sigma^2 - \theta_0}e^{-2\rho(\theta/\sigma - \theta_0)} - \sqrt{2}\theta_0 e^{\theta_0\rho - \sqrt{2}\theta/\sigma} + (1 + \sqrt{2}\theta_0 e^{\theta_0\rho} - e^{2\theta_0\rho})e^{-2\rho\theta/\sigma}$$

$$+ 4\theta\theta_0 \left[\left(1 - \frac{1}{\sigma^2}\right)\int_0^\infty (E_0 R_x - \rho)dx + \frac{1}{\sigma}\int_0^\infty (E_0 R_{x/\sigma} - \rho)d(E_0 R_x - \rho)\right]$$

$$+ o(\theta_0\theta).$$

The third lemma approximates the expected length of time until the CUSUM process goes back to zero after the change has occurred. The proof is similar to that of Lemma 6.2 and is thus omitted.

Lemma 6.3: *As $\theta_0, \theta \to 0$ at the same order,*

$$E[\tau_{-M}; \tau_{-M} = \infty]$$

$$= -\frac{2\theta_0\sigma}{\theta}e^{-2(\theta/\sigma - \theta_0)\rho}\left(\frac{1}{4\sigma(\theta/\sigma^2 - \theta_0)^2} + \frac{\rho}{2(\theta/\sigma^2 - \theta_0)}\right) - \frac{\theta_0\sigma}{\theta}$$

$$- \frac{2\theta_0\sigma^2\rho}{\theta}\left[\left(1 - \frac{1}{\sigma^2}\right)\int_0^\infty (E_0 R_x - \rho)dx + \frac{2\theta_0}{\theta\sigma}\int_0^\infty E_0 R_{x/\sigma} d(E_0 R_x - \rho)\right] + o(1).$$

The next lemma gives the approximation to the bias of the estimator when the CUSUM process directly crosses the threshold without going back to zero, which is the maximum point of S_n' with maximum value M if we look at the CUSUM process backward in time starting from the change-point. The technique extends Lemma 2.3 of Wu (2004), and a sketch of the proof is provided in the appendix.

Lemma 6.4: *As $\theta_0, \theta \to 0$ at the same order,*

$$E[\eta_M; \tau_{-M} = \infty] = \frac{1}{2\theta_0^2} - \frac{1}{2(\theta/\sigma^2 - \theta_0)^2}e^{-\frac{2\theta}{\sigma^2}\rho(\sigma - 1)} + o(1).$$

Finally, by summing up the results of Lemmas $6.1 - 6.4$, we have the following second-order approximation for the bias.

Theorem 6.2: *As $\theta_0, \theta \to 0$ at the same order, the asymptotic bias of $\hat{\nu}$ is given by*

$$\lim_{\nu\to\infty}\lim_{d\to\infty} E^\nu[\hat{\nu} - \nu|N > \nu] = -\frac{\theta_0}{2(\theta/\sigma)^2(\theta/\sigma^2 - \theta_0)}e^{-2\rho(\theta/\sigma - \theta_0)}$$

$$- \frac{2\theta_0\sigma}{\theta}e^{-2\rho(\theta/\sigma - \theta_0)}\left(\frac{1}{4\sigma(\theta/\sigma^2 - \theta_0)^2} + \frac{\rho}{2(\theta/\sigma^2 - \theta_0)}\right)$$

$$+ \frac{\theta_0}{4(\theta/\sigma^2 - \theta_0)} - \frac{\sqrt{2}\theta_0}{2(\theta/\sigma)^2}e^{\theta_0\rho} + \frac{e^{-2\rho\theta/\sigma}}{2(\theta/\sigma)^2}(1 + \sqrt{2}\theta_0 e^{\theta_0\rho} - e^{2\theta_0\rho})$$

$$- \frac{2\theta_0\sigma}{\theta}\left(\rho - \frac{1}{\sqrt{2}}\right)^2 - \frac{1}{2\theta_0^2} + \frac{1}{2(\theta/\sigma^2 - \theta_0)^2}e^{2\theta - \rho(1-\sigma)/\sigma^2} + o(1).$$

The second-order approximation for the absolute bias can be obtained similarly.

Remark: When $\theta = -\theta_0$, the first-order bias of $\hat{\nu}$ after some algebraic calculation can be simplified as

$$E^\nu[\hat{\nu} - \nu | N > \nu] \approx \frac{(\sigma^2 - 1)(\sigma^4 + 3\sigma^2 + 1)}{2\theta^2(1 + \sigma^2)^2} - \frac{2\rho(\sigma - 1)}{\theta(1 + \sigma^2)}$$

$$- \frac{\sigma\rho^2(\sigma^3 - \sigma^2 - 1)(\sigma^2 - 1)}{(1 + \sigma^2)^2} - \frac{\sigma\rho(\sigma - 1)}{\sqrt{2}} - \frac{\sigma^2}{4(1 + \sigma^2)} + o(1). \tag{6.1}$$

If further, $\sigma = 1$, we have

$$E^\nu[\hat{\nu} - \nu | N > \nu] \approx -\frac{1}{8} + o(1),$$

as given in Chapter 2. Thus, the bias is very sensitive to the post-change variance σ^2. A numerical demonstrations can be seen in Section 6.5.

6.3 Bias of $\hat{\theta}$ and $\hat{\sigma}^2$

In this section, we extend the results of Chapter 4 where the case $\sigma^2 = 1$ is considered. We first consider the bias and leave the construction of the confidence interval to Section 6.4. The approach here closely follows the lines of Chapter 4.

6.3.1 Bias of $\hat{\theta}$

For simplicity, we assume all related Taylor expansions are valid as well as under derivatives. First, we define

$$N_x = \inf\{n > 0 : S_n \le 0 \text{ or } > d\}$$

for $S_0 = x \ge 0$. As in Chapter 4 and Section 6.2, as $\nu \to \infty$ and d approaches infinity, we note that if after the change-point, the CUSUM process comes back to zero, or equivalently $S_{N_M} \le 0$, then $\hat{\theta}$ is the sample mean for a random walk staying positive until it crosses the threshold d. On the other hand, if $S_{N_M} > d$, then $\hat{\theta}$ is the sample mean with sample size $N_M + \eta_M$, which is the time to cross the threshold after the change, plus the time before the change. Thus, we have

$$\begin{aligned} E^\nu[\hat{\theta} - \theta | N > \nu] &= E^\nu[\hat{\theta} - \theta; S_{N_M} \le 0 | N > \nu] + E^\nu[\hat{\theta} - \theta; S_{N_M} > d | N > \nu] \\ &= E_{(\theta,\sigma)}\left[\frac{S_{N_0}}{N_0} - \theta | S_{N_0} > d\right] P(\tau_{-M} < \infty)(1 + o(1)) \\ &\quad + E\left[\frac{S_{N_M}}{N_M + \eta_M}; S_{N_M} > d\right](1 + o(1)), \end{aligned} \tag{6.2}$$

as
$$P(S_{N_M} > d) = P(\tau_{-M} = \infty)(1 + o(1)).$$

For the convenience of notation, we use the conditional expectation in the first term.

In the following, we shall give the first-order result for the related quantities as $d \to \infty$.

The first lemma gives the bias for the sample mean for a random walk staying positive until it crosses d.

Lemma 6.5: *As $d \to \infty$,*

$$E_{(\theta,\sigma)}\left(\frac{S_{N_0}}{N_0} - \theta | S_{N_0} > d\right) = \frac{\sigma^2}{d} \frac{\frac{\partial}{\partial \theta}(\theta P_{\theta/\sigma}(\tau_- = \infty))}{P_{\theta/\sigma}(\tau_- = \infty)}(1 + o(1)).$$

Proof: By using Wald's likelihood ratio identity by changing parameter θ to 0 and the property of normal density, we have

$$\frac{\partial}{\partial \theta} E_{(\theta,\sigma)}\left[\frac{1}{N_0}; S_{N_0} > d\right] = \frac{\partial}{\partial \theta} E_{(0,\sigma)}\left[\frac{1}{N_0} exp\left(\frac{1}{\sigma^2}\left(S_{N_0} - N\frac{\theta^2}{2}\right)\right); S_{N_0} > d\right]$$

$$= \frac{1}{\sigma^2} E_{(0,\sigma)}\left[\frac{S_{N_0} - N_0\theta}{N_0} exp\left(\frac{1}{\sigma^2}\left(S_{N_0} - N\frac{\theta^2}{2}\right)\right); S_{N_0} > d\right]$$

$$= \frac{1}{\sigma^2} E_{(\theta,\sigma)}\left[\frac{S_{N_0}}{N_0} - \theta; S_{N_0} > d\right].$$

Thus,

$$E_{(\theta,\sigma)}\left[\frac{S_{N_0}}{N_0} - \theta; S_{N_0} > d\right] = \sigma^2 \frac{\partial}{\partial \theta} E_{(\theta,\sigma)}\left[\frac{1}{N_0}; S_{N_0} > d\right].$$

As $d \to \infty$,

$$N_0 = \frac{d}{\theta}(1 + o_p(1)),$$

given $S_{N_0} > d$, where $o_p(1)$ represents $o(1)$ in probability and

$$P_{(\theta,\sigma)}(S_{N_0} > d) = P_{(\theta,\sigma)}(\tau_- = \infty)(1 + o(1))$$
$$= P_{\theta/\sigma}(\tau_- = \infty)(1 + o(1)).$$

By passing the limit inside the partial derivative, we get

$$E_{(\theta,\sigma)}\left[\frac{S_{N_0}}{N_0} - \theta; S_{N_0} > d\right] = \frac{1}{d}\frac{\partial}{\partial \theta}(\theta P_{\theta/\sigma}(\tau_- = \infty))(1 + o(1)),$$

which is equivalent to the expected result by writing into the conditional expected version.

The next lemma gives the result for the second term in (6.2).

Lemma 6.6: *As* $d \to \infty$,

$$E\left[\frac{S_{N_M}}{N_M + \eta_M} - \theta; S_{N_M} > d\right]$$

$$= \frac{1}{d}\left[\sigma^2 \frac{\partial}{\partial\theta}(\theta P(\tau_{-M} = \infty)) + \theta E[(M - \theta\eta_M); \tau_{-M} = \infty]\right].$$

Proof: We first write

$$E\left[\frac{S_{N_M}}{N_M + \eta_M} - \theta; S_{N_M} > d\right]$$

$$= E\left[\frac{S_{N_M} - M}{N_M + \eta_M} - \theta; S_{N_M} > d\right] + E\left[\frac{M}{N_M + \eta_M}; S_{N_M} > d\right]$$

$$= E\left[\frac{S_{N_M} - M}{N_M} - \theta; S_{N_M} > d\right]$$

$$+ E\left[\frac{1}{N_M}\left(M - \frac{S_{N_M} - M}{N_M}\eta_M\right); S_{N_M} > d\right](1 + o(1)). \qquad (6.3)$$

The first term of (6.3) can be evaluated as in Lemma 6.1 by using Wald's likelihood ratio identity and the normal property with $S_0 = M$. We have

$$E\left[\frac{S_{N_M} - M}{N_M} - \theta; S_{N_M} > d\right] = \frac{\sigma^2}{d}\frac{\partial}{\partial\theta}(\theta P(\tau_{-M} = \infty))(1 + o(1)).$$

For the second term of (6.3), we note that

$$(S_{N_M} - M)/N_M = \theta(1 + o_p(1)),$$

given $S_{N_M} > d$. The result follows by combining the two approximations. $\quad\square$

Combining Lemmas 6.5 and 6.6, we get the following first-order asymptotic bias for $\hat\theta$.

Theorem 6.3: *As* $d, \nu \to \infty$,

$$E^\nu[\hat\theta - \theta | N > \nu] = \frac{\sigma^2}{d}\{1 + \theta\frac{\partial}{\partial\theta}P_{\theta/\sigma}(\tau_- = \infty)\frac{P(\tau_{-M} < \infty)}{P_{\theta/\sigma}(\tau_- = \infty)}$$

$$+ \theta\frac{\partial}{\partial\theta}P(\tau_{-M} = \infty) + \frac{\theta}{\sigma^2}E[(M - \theta\eta_M); \tau_{-M} = \infty]\}(1 + o(1)).$$

Now, we consider the local expansion when both θ_0 and θ are small. From Corollary 4.1 of Chapter 4, we have

$$\theta\frac{\frac{\partial}{\partial\theta}P_{\theta/\sigma}(\tau_- = \infty)}{P_{\theta/\sigma}(\tau_- = \infty)} = 1 + o(1);$$

and from Lemma 6.2,

$$P(\tau_{-M} < \infty) = -\frac{\theta_0}{\theta/\sigma^2 - \theta_0}(1 + o(1)).$$

On the other hand, as $\theta_0 \to 0$ from the renewal theorem, as shown in Chapter 4,

$$-2\theta_0 M \to Y,$$

in distribution, where Y is a standard exponential random variable. Thus,

$$
\begin{aligned}
E[M; \tau_{-M} = \infty] &= E_{\theta_0}[M P_{\theta/\sigma}(\tau_{-M/\sigma} = \infty)] \\
&= -\frac{1}{2\theta_0} E\left[Y\left(1 - exp\left(\frac{\theta}{\theta_0\sigma^2}Y\right)\right)\right](1 + o(1)) \\
&= -\frac{1}{2\theta_0}\left(1 - \frac{\theta_0^2}{(\theta/\sigma^2 - \theta_0)^2}\right)(1 + o(1)).
\end{aligned}
$$

From Lemma 6.4, we have

$$E[\eta_M; \tau_{-M} = \infty] = \left(\frac{1}{2\theta_0^2} - \frac{1}{2(\theta/\sigma^2 - \theta_0)^2}\right)(1 + o(1)).$$

Combining the above approximations, we have the following local approximation.

Corollary 6.1: *As $d, \nu \to \infty$, locally,*

$$E^\nu[\hat{\theta} - \theta | N > \nu] = \frac{\sigma^2}{d}\left(2 - \frac{\theta/\sigma^2}{(\theta/\sigma^2 - \theta_0)} + \frac{\theta^2 - \theta\theta_0}{2\sigma^2(\theta/\sigma^2 - \theta_0)^2} - \frac{\theta^2 + \theta\theta_0}{2\theta_0^2\sigma^2}\right)$$

$$\times (1 + o(1)).$$

When $\theta = -\theta_0$,

$$E^\nu[\hat{\theta} - \theta | N > \nu] = \frac{\sigma^2}{d}\left(2 - \frac{1}{(1 + \sigma^2)^2}\right)(1 + o(1)). \qquad (6.4)$$

Remark: The case $\sigma^2 = 1$ is studied in Chapter 4, which shows that if $\theta = -\theta_0$,

$$E^\nu[\hat{\theta} - \theta | N > \nu] = \frac{7}{4d}(1 + o(1)).$$

One can see that larger σ's enhance the bias of $\hat{\theta}$. A numerical demonstration can be seen in Section 6.5.

6.3.2 Bias of $\hat{\sigma}^2$

Next, we study the bias of $\hat{\sigma}^2$, which is also needed to construct the confidence interval for θ. The technique used here is a combination of the ones of Woodroofe (1992), Coad and Woodroofe (1998), and Chapter 4. We directly consider the quasistationary state case. Denoting $\{X_i'\}$ as an independent copy of $\{X_i\}$, we have

$$E^\nu[\hat{\sigma}^2 - \sigma^2|N > \nu] = E\left[\frac{1}{N_0}\sum_{i=1}^{N_0}\left(X_i - \frac{S_{N_0}}{N_0}\right)^2 - \sigma^2|S_{N_0} > d\right] P(\tau_{-M} < \infty)$$

$$+ E[\frac{1}{N_M + \eta_M}[\sum_{i=1}^{\eta_M}\left(X_i' - \frac{S_{N_M}}{N_M + \eta_M}\right)^2$$

$$+ \sum_{i=1}^{N_M}\left(X_i - \frac{S_{N_M}}{N_M + \eta_M}\right)^2] - \sigma^2; S_{N_M} > d]. \tag{6.5}$$

The first term of (6.5) is the bias of sample variance if the CUSUM process goes back to zero after the change-point, and the second term of (6.5) is the sample variance if the CUSUM process directly crosses the threshold after the change-point, which consists of two parts for the total sum of the errors: One part is the sum before the change-point and the other part is the sum after the change.

For the first term of (6.5), we use Wald's likelihood ratio identity by changing parameter (θ, σ) to $(\theta, 1)$, and using the normal density property, and get

$$\frac{\partial}{\partial \sigma^2}E_{(\theta,\sigma)}\left[\frac{1}{N_0}; S_{N_0} > d\right]$$

$$= \frac{\partial}{\partial \sigma^2}E_{(\theta,1)}\left[\frac{1}{N_0\sigma^2}exp\left(-\frac{1}{2}\left(\frac{1}{\sigma^2} - 1\right)\sum_{i=1}^{N_0}(X_i - \theta)^2\right); S_{N_0} > d\right]$$

$$= \frac{1}{2\sigma^4}$$

$$\times E_{(\theta,1)}\left[\frac{1}{N_0\sigma^2}\sum_{i=1}^{N_0}(X_i - \theta)^2 exp\left(-\frac{1}{2}\frac{1}{\sigma^2} - 1)\sum_{i=1}^{N_0}(X_i - \theta)^2\right); S_{N_0} > d\right]$$

$$- \frac{1}{\sigma^2}E_{(\theta,1)}\left[\frac{1}{N_0\sigma^2}exp\left(-\frac{1}{2}\left(\frac{1}{\sigma^2} - 1\right)\sum_{i=1}^{N_0}(X_i - \theta)^2\right); S_{N_0} > d\right]$$

$$= \frac{1}{2\sigma^4}\{E_{(\theta,\sigma)}\left[\frac{1}{N_0}\sum_{i=1}^{N_0}\left(X_i - \frac{S_{N_0}}{N_0}\right)^2; S_{N_0} > d\right]$$

$$+ E_{(\theta,\sigma)}\left[\left(\frac{S_{N_0}}{N_0} - \theta\right)^2; S_{N_0} > d\right] - \sigma^2 P_{(\theta,\sigma)}(S_{N_0} > d)\}.$$

Thus,

$$E\left[\frac{1}{N_0}\sum_{i=1}^{N_0}\left(X_i - \frac{S_{N_0}}{N_0}\right)^2 - \sigma^2; S_{N_0} > d\right]$$

$$= 2\sigma^4\frac{\partial}{\partial\sigma^2}E_{(\theta,\sigma)}\left[\frac{1}{N_0}; S_{N_0} > d\right] - E_{(\theta,\sigma)}\left[\left(\frac{S_{N_0}}{N_0} - \theta\right)^2; S_{N_0} > d\right]$$

$$= \frac{2\sigma^4}{d}\frac{\partial}{\partial\sigma^2}(\theta P_{(\theta,\sigma)}(\tau_- = \infty))\frac{\theta}{d}\sigma^2 P_{(\theta,\sigma)}(\tau_- = \infty)(1 + o(1))$$

$$= \frac{2\theta\sigma^4}{d}\left[\frac{\partial}{\partial\sigma^2}P_{\theta/\sigma}(\tau_- = \infty) - \frac{1}{2\sigma^2}P_{\theta/\sigma}(\tau_- = \infty)\right](1 + o(1)),$$

where we use the fact that

$$\frac{1}{N_0}\sum_{i=1}^{N_0}\left(X_i - \frac{S_{N_0}}{N_0}\right)^2 = \sigma^2(1 + o_p(1)),$$

given $S_{N_0} > d$.

For the second term of (6.5), we note that N_M dominates η_M and

$$N_M + \eta_M = N_M\left(1 + \frac{\eta_M}{N_M}\right).$$

Up to the first-order, we can replace S_{N_M} with $S_{N_M} - M$ and $N_M + \eta_M$ with N_M for the first summation. Thus, we have

$$E\left[\frac{1}{N_M + \eta_M}\left[\sum_{i=1}^{\eta_M}(X_i' - \frac{S_{N_M}}{N_M + \eta_M})^2 + \sum_{i=1}^{N_M}(X_i - \frac{S_{N_M}}{N_M + \eta_M})^2\right] - \sigma^2; S_{N_M} > d\right]$$

$$= E\left[\frac{1}{N_M}\sum_{i=1}^{\eta_M}(X_i' - \theta)^2; S_{N_M} > d\right]$$

$$+ E\left[\frac{1}{N_M}\sum_{i=1}^{N_M}\left(X_i - \frac{S_{N_M} - M}{N_M}\right)^2 - \sigma^2; S_{N_M} > d\right]$$

$$- E\left[\frac{\eta_M}{N_M}\frac{1}{N_M}\sum_{i=1}^{N_M}\left(X_i - \frac{S_{N_M} - M}{N_M}\right)^2; S_{N_M} > d\right](1 + o(1))$$

$$= \frac{\theta}{d}E\left[\sum_{i=1}^{\eta_M}[(X_i' - \theta)^2 - \sigma^2]; \tau_{-M} = \infty\right] + \frac{2\theta\sigma^4}{d}\frac{\partial}{\partial\sigma^2}P(\tau_{-M} = \infty)$$

$$- \frac{\theta\sigma^2}{d}P(\tau_{-M} = \infty)(1 + o(1)),$$

where in the final step, we use the similar technique for the first term and the first summation combines the first sum and the third sum in the second equation from the last since

$$\frac{1}{N_M} \sum_{i=1}^{N_M} \left(X_i - \frac{S_{N_M} - M}{N_M} \right)^2 = \sigma^2 (1 + o_p(1)).$$

Finally, we have the following first-order bias for $\hat{\sigma}^2$.

Theorem 6.4: *As $d, \nu \to \infty$,*

$$E^\nu[\hat{\sigma}^2 - \sigma^2 | N > \nu] = \frac{2\theta\sigma^4}{d} \frac{\partial}{\partial\sigma^2} P_{\theta/\sigma}(\tau_- = \infty) \frac{P(\tau_{-M} < \infty)}{P_{\theta/\sigma}(\tau_- = \infty)} - \frac{\theta\sigma^2}{d}$$

$$+ \frac{2\theta\sigma^4}{d} \frac{\partial}{\partial\sigma^2} P(\tau_{-M} = \infty) + \frac{\theta}{d} E\left[\sum_{i=1}^{\eta_M} ((X_i' - \theta)^2 - \sigma^2); \tau_{-M} = \infty \right] (1 + o(1)).$$

The local evaluation is technically complicated and will not be pursued in this work.

6.4 Corrected Confidence Interval

In this section, we further extend the result by constructing a corrected confidence interval for θ. The basic idea is to correct the asymptotic normal pivot

$$\sqrt{N - \hat{\nu}}(\hat{\theta} - \theta)/\hat{\sigma}$$

by deriving its first-order bias and variance, respectively, as in Chapter 4. As the numerical approximation is more complicated and we give only the asymptotic first-order result.

The first theorem gives the bias of the asymptotic pivot, and its proof is similar to that of Theorem 4.2.

Theorem 6.5: *As $d, \nu \to \infty$,*

$$E^\nu[\sqrt{N - \hat{\nu}}(\hat{\theta} - \theta)/\hat{\sigma}|N > v] = \frac{a(\theta, \sigma^2)}{\sqrt{d}}(1 + o(1)),$$

where

$$a(\theta, \sigma^2) = \frac{1}{\sigma} \frac{\partial}{\partial\theta} (\sqrt{\theta} P_{\theta/\sigma}(\tau_- = \infty)) \frac{P(\tau_{-M} < \infty)}{P_{\theta/\sigma}(\tau_- = \infty)}$$

$$+ \frac{1}{\sigma} \frac{\partial}{\partial\theta} (\sqrt{\theta} P(\tau_{-M} = \infty)) + \frac{\sqrt{\theta}}{\sigma} E[M - \theta\eta_M; \tau_{-M} = \infty].$$

The main difficulty comes from the evaluation for the variance of Z_N up to the order $O(1/d)$ after correcting the estimated bias. We first write

$$E^\nu\left[\left(\frac{\sqrt{N-\hat\nu}(\hat\theta-\theta)}{\hat\sigma}-\frac{a(\hat\theta,\hat\sigma^2)}{\sqrt{d}}\right)^2|N>\nu\right]=E^\nu\left[\frac{(N-\hat\nu)(\hat\theta-\theta)^2}{\hat\sigma^2}|N>\nu\right]$$

$$-\frac{2}{\sqrt{d}}E\left[\frac{\sqrt{N-\hat\nu}(\hat\theta-\theta)}{\hat\sigma}a(\hat\theta,\hat\sigma^2)|N>\nu\right]+\frac{1}{d}a(\theta,\sigma^2)(1+o(1)).$$

In the second term above, we use the Taylor expansion in the second term for $a(\hat\theta,\hat\sigma^2)$ at (θ,σ^2) and note that $\hat\theta$ and $\hat\sigma$ are asymptotically uncorrelated. Thus, we have

$$E^\nu\left[\frac{\sqrt{N-\hat\nu}(\hat\theta-\theta)}{\hat\sigma}a(\hat\theta,\hat\sigma^2)|N>\nu\right]=a(\theta,\sigma^2)E^\nu\left[\frac{\sqrt{N-\hat\nu}(\hat\theta-\theta)}{\hat\sigma}|N>\nu\right]$$

$$+\frac{\partial}{\partial\theta}a(\theta,\sigma^2)E^\nu\left[\frac{\sqrt{N-\hat\nu}(\hat\theta-\theta)^2}{\hat\sigma}|N>\nu\right](1+o(1))$$

$$=\frac{a^2(\theta,\sigma^2)}{\sqrt{d}}+\frac{\partial}{\partial\theta}a(\theta,\sigma^2)\frac{\sqrt{\theta}\sigma}{\sqrt{d}}(1+o(1)),$$

where in the last equation, we use the fact

$$E^\nu\left[\frac{(N-\hat\nu)(\hat\theta-\theta)^2}{\sigma}|N>\nu\right]=1+o(1),$$

by the asymptotic normality of Z_N.

Thus, we have

$$E^\nu\left[\left(\frac{\sqrt{N-\hat\nu}(\hat\theta-\theta)}{\hat\sigma}-\frac{a(\hat\theta,\hat\sigma^2)}{\sqrt{d}}\right)^2|N>\nu\right]$$

$$=E^\nu\left[\frac{(N-\hat\nu)(\hat\theta-\theta)^2}{\hat\sigma^2}|N>\nu\right]-\frac{2\sigma\sqrt{\theta}}{d}\frac{\partial}{\partial\theta}a(\theta,\sigma^2)-\frac{a^2(\theta,\sigma^2)}{d}(1+o(1)).$$

The next theorem gives the first-order approximation for the second moment of the asymptotic pivot.

Theorem 6.6: *As $d,\nu\to\infty$,*

$$E^\nu\left[\frac{(N-\hat\nu)(\hat\theta-\theta)^2}{\hat\sigma^2}|N>\nu\right]=1+\frac{b(\theta,\sigma^2)}{d}(1+o(1)),$$

where

$$b(\theta, \sigma^2) = P(\tau_{-M} < \infty)$$

$$\times \left(\theta + \sigma^2 \frac{\frac{\partial^2}{\partial^2 \theta}(\theta P_{\theta/\sigma}(\tau_- = \infty))}{P_{\theta/\sigma}(\tau_- = \infty)} - 2\sigma^2 \theta \frac{\frac{\partial}{\partial \sigma^2} P_{\theta/\sigma}(\tau_- = \infty)}{P_{\theta/\sigma}(\tau_- = \infty)} \right)$$

$$+ \sigma^2 \frac{\partial^2}{\partial^2 \theta}(\theta P(\tau_{-M} = \infty)) - \frac{1}{\sigma^2}\left\{ \theta E\left[\sum_{i=1}^{\eta_M}((X_i' - \theta)^2 - \sigma^2); \tau_{-M} = \infty \right] \right.$$

$$\left. + 2\sigma^2 \frac{\partial}{\partial \sigma^2}(\theta P(\tau_{-M} = \infty)) - \theta\sigma^2 P(\tau_{-M} = \infty) \right\}$$

$$+ 2\frac{\partial}{\partial \theta} E[\theta(M - \theta\eta_M); \tau_{-M} = \infty]$$

$$+ \frac{\theta}{\sigma^2} E[(M - \theta\eta_M)^2; \tau_{-M} = \infty] + \theta E[\eta_M; \tau_{-M} = \infty].$$

Proof: By separating the case of $S_{N_M} \leq 0$ or $S_{N_M} > d$, we first write

$$E^\nu\left[\frac{(N - \hat{\nu})(\hat{\theta} - \theta)^2}{\hat{\sigma}^2} | N > \nu \right] = E_{(\theta,\sigma)}\left[\frac{N_0(\hat{\theta}_0 - \theta)^2}{\hat{\sigma}_0^2} | S_{N_0} > d \right] P(\tau_{-M} < \infty)$$

$$+ E\left[\frac{(N_M + \eta_M)(\hat{\theta}_M - \theta)^2}{\hat{\sigma}_M^2}; S_{N_M} > d \right](1 + o(1)),$$

where

$$\hat{\theta}_0 = S_{N_0}/N_0$$

and

$$\hat{\sigma}_0^2 = \frac{1}{N_0} \sum_{i=1}^{N_0}(X_i - \hat{\theta}_0)^2;$$

and $\hat{\theta}_M$ and $\hat{\sigma}_M^2$ are defined accordingly.

By using Wald's likelihood ratio identity as in Section 3.2 by changing parameter (θ, σ) to $(\theta, 1)$ and taking the derivative twice on $E_{(\theta,\sigma)}\left[\frac{1}{N_0\hat{\sigma}^2}; S_{N_0} > d \right]$ with respect to σ^2, we have

$$E_{(\theta,\sigma)}\left[\frac{N_0(\hat{\theta}_0 - \theta)^2}{\hat{\sigma}_0^2}; S_{N_0} > d \right]$$

$$= \sigma^4 \frac{\partial^2}{\partial^2 \theta} E_{(\theta,\sigma)}\left[\frac{1}{N_0\hat{\sigma}_0^2}; S_{N_0} > d \right] + \sigma^2 E_{(\theta,\sigma)}\left[\frac{1}{\hat{\sigma}_0^2}; S_{N_0} > d \right]$$

$$= \frac{\sigma^4}{d} \frac{\partial^2}{\partial^2 \theta}(\theta P_{\theta/\sigma}(\tau_- = \infty)) + \sigma^2 E_{(\theta,\sigma)}\left[\frac{1}{\hat{\sigma}_0^2}; S_{N_0} > d \right](1 + o(1)).$$

Now, by the Taylor expansion and using part of the results in Theorem 3.2, we have

$$\sigma^2 E_{(\theta,\sigma)}\left[\frac{1}{\hat{\sigma}_0^2}; S_{N_0} > d \right] = P_{(\theta,\sigma)}(\tau_- = \infty) - \frac{1}{\sigma^2} E_{(\theta,\sigma)}[\hat{\sigma}_0^2 - \sigma^2; S_{N_0} > d](1 + o(1))$$

$$= P_{\theta/\sigma}(\tau_- = \infty) - \frac{2\sigma^2\theta}{d}\left(\frac{\partial}{\partial\sigma^2}P_{\theta/\sigma}(\tau_- = \infty) - \frac{1}{2\sigma^2}P_{\theta/\sigma}(\tau_- = \infty)\right)(1 + o(1)).$$

Thus,

$$E\left[\frac{N_0(\hat{\theta}_0 - \theta)^2}{\hat{\sigma}_0^2}|S_{N_0} > d\right] = 1 + \frac{1}{d}\{\theta + \sigma^2\frac{1}{P_{\theta/\sigma}(\tau_- = \infty)}\frac{\partial^2}{\partial^2\theta}(\theta P_{\theta/\sigma}(\tau_- = \infty))$$

$$- 2\sigma^2\theta\frac{1}{P_{\theta/\sigma}(\tau_- = \infty)}\frac{\partial}{\partial\sigma^2}P_{\theta/\sigma}(\tau_- = \infty)\}(1 + o(1)).$$

Similarly, we have

$$E\left[\frac{(N_M+\eta_M)(\hat{\theta}_M-\theta)^2}{\hat{\sigma}_M^2}; S_{N_M} > d\right]$$

$$= E\left[\frac{N_M\left(\frac{S_{N_M}-M}{N_M} - \theta + \frac{1}{M}(M - \theta\eta_M)\right)^2}{\hat{\sigma}_M^2}; S_{N_M} > d\right]$$

$$+ \frac{\theta}{d}E[\eta_M; \tau_{-M} = \infty](1 + o(1))$$

$$= E\left[\frac{N_M\left(\frac{S_{N_M}-M}{N_M} - \theta\right)^2}{\hat{\sigma}_M^2}; S_{N_M} > d\right]$$

$$+ 2E\left[\frac{\left(\frac{S_{N_M}-M}{N_M} - \theta\right)(M - \theta\eta_M)}{\hat{\sigma}_M^2}; S_{N_M} > d\right]$$

$$+ E\left[\frac{\frac{1}{N_M}(M - \theta\eta_M)^2}{\hat{\sigma}_M^2}; S_{N_M} > d\right] + \frac{\theta}{d}E[\eta_M; \tau_{-M} = \infty](1 + o(1))$$

$$= \sigma^2E\left[\frac{1}{\hat{\sigma}_M^2}; S_{N_M} > d\right] + \sigma^4\frac{\partial^2}{\partial^2\theta}\left[\frac{1}{N_M\hat{\sigma}_M^2}; S_{N_M} > d\right]$$

$$+ 2\sigma^2\frac{\partial}{\partial\theta}E\left[\frac{M - \theta\eta_M}{N_M\hat{\sigma}_M^2}; S_{N_M} > d\right]$$

$$+ \frac{\theta}{d\sigma^2}E[(M - \theta\eta_M)^2; \tau_{-M} = \infty] + \frac{\theta}{d}E[\eta_M; \tau_{-M} = \infty](1 + o(1))$$

$$= P(\tau_{-M} = \infty) - \frac{1}{\sigma^2}E[\hat{\sigma}_M^2 - \sigma^2; S_{N_M} > d]$$

$$+ \frac{\sigma^2}{d}\frac{\partial^2}{\partial^2\theta}(\theta P(\tau_{-M} = \infty)) + \frac{2}{d}\frac{\partial}{\partial\theta}E[\theta(M - \theta\eta_M); \tau_{-M} = \infty]$$

$$+ \frac{\theta}{d\sigma^2}E[(M - \theta\eta_M)^2; \tau_{-M} = \infty] + \frac{\theta}{d}E[\eta_M; \tau_{-M} = \infty](1 + o(1)).$$

Using Theorem 3.1, we get the result.

Theoretically, a corrected asymptotic normal pivot can be formed as

$$\frac{\left(\sqrt{N-\hat{\nu}}\frac{\hat{\theta}-\theta}{\hat{\sigma}} - \frac{a(\hat{\theta},\hat{\sigma}^2)}{\sqrt{d}}\right)}{\left(1 + \frac{1}{d}\left(b(\hat{\theta},\hat{\sigma}^2) - 2\sqrt{\theta}\frac{\partial}{\partial\theta}a(\hat{\theta},\hat{\sigma}^2) - a^2(\hat{\theta},\hat{\sigma}^2)\right)\right)^{1/2}}.$$

Heuristically, we can argue by following the very weak expansion technique as demonstrated in Woodroofe and Coad (1998) and Woodroofe (1986) that the above approximation is accurate up to the order $O(d^{-3/2})$.

An alternative approximations such as Edgeworth expansion and saddle-point approximation can be used to obtain higher-order approximations. In the sequential sampling plan case, theses techniques are very difficult to apply even in some simple situations. Although the above corrected normal pivot does not correct the shape of the distribution, it is general enough to give a second-order approximation, particularly in this change-point problem where the conventional method seems hard to apply.

6.5   Numerical Evaluation

In this section, we conduct a simulation study to check the accuracy of the approximations. Due to the complicated form for the confidence interval, our evaluation is restricted to the biases of the change-point and post-change mean estimators, which, at least, gives some numerical supports for a reliable bias correction method.

Table 6.1 gives a comparison between the simulated and approximated values for the biases of $\hat{\nu}$ and $\hat{\theta}$ respectively. For $\theta = -\theta_0 = 0.25$ and 0.5, $d = 10.0$ and $\sigma = 1.0, 1.05$ and 1.1, 1000 simulations are conducted for $\nu = 0, 50, 75$, and 100, respectively, on the Dell Workstation by using the PC version 6.1 of S-Plus. Only those trials with $N > \nu$ are used to evaluate the biases. The simulated values with standard errors in the parentheses are reported followed by the approximated values in the bottom cell. The approximated values are obtained by using (6.1) and (6.4), respectively. Listed also are the corresponding probabilities of $P^\nu(N > \nu)$.

By looking at the table, we can make several comments. First, even though the approximations are given at the first-order that are quite accurate, particularly when $\theta = 0.25$. Second, the effect of ν on the bias of $\hat{\theta}$ is quite small, while it is significant on the bias of $\hat{\nu}$, particularly when $\nu = 0$. Third, there is little difference between the cases $\nu = 50, 75$, and 100, which shows that the convergence is extremely fast. We also note that since the variance of the change-point estimator is large, one may expect larger standard errors for the bias estimator. However, the standard error for the post-change mean estimator is quite small, which makes a bias correction possible.

Table 6.1: Biases of $\hat{\nu}$ and $\hat{\theta}$

$\theta = -\theta_0$	σ	ν	$P(N > \nu)$	$E[\hat{\nu} - \nu | N > \nu]$	$E[\hat{\theta} - \theta | N > \nu]$
0.25	1.0	100	0.957	$-0.022(0.547)$	$0.157(0.007\ )$
		75	0.966	$1.239(0.561)$	$0.172(0.008)$
		50	0.990	$0.781(0.543)$	$0.176(0.0071)$
		0	1.000	$7.74(0.420)$	$0.206(0.008)$
		approx.		-0.125	0.175
	1.05	100	0.963	$1.107(0.575)$	$0.191(0.008)$
		75	0.981	$2.145(0.620)$	$0.191\ (0.007)$
		50	0.992	$1.454(0.577)$	$0.194(0.0075)$
		0	1.000	$7.61(0.400)$	$0.227(0.008)$
		approx.		1.024	0.196
	1.1	100	0.965	$2.223(0.634)$	$0.212(0.008)$
		75	0.979	$2.428(0.583)$	$0.215(0.008)$
		50	0.991	$2.725(0.560)$	$0.222(0.008)$
		0	1.000	$8.59(0.440)$	$0.267(0.009)$
		approx.		2.096	0.217
0.5	1.0	100	0.999	$-0.101(0.159)$	$0.141(0.008)$
		75	1.000	$0.244(0.166)$	$0.155\ (0.009)$
		50	1.000	$-0.180(0.171)$	$0.144(0.008)$
		0	1.0	$1.496(0.091)$	$0.175(0.009\ )$
		approx.		-0.125	0.175
	1.05	100	1.000	$0.031(0.180)$	$0.162(0.009)$
		75	1.000	$0.244(0.166)$	$0.155(0.009)$
		50	1.000	$0.294(0.196)$	$0.159(0.009)$
		0	1.000	$1.845(0.131)$	$0.179(0.009)$
		approx.		0.256	0.196
	1.10	100	1.000	$0.544(0.197)$	$0.175(0.009)$
		75	0.997	$0.449(0.181)$	$0.163(0.009)$
		50	1.000	$0.588(0.171)$	$0.200(0.010)$
		0	1.000	$2.15(0.134)$	$0.203(0.009)$
		approx.		0.524	0.217

6.6 Appendix

6.6.1 Proof of Lemma 6.2

By separating the event $\{M > 0\}$ from the event $\{M = 0\}$, which is equivalent to $\{\tau_+ = \infty\}$, we first write

$$P(\tau_{-M} < \infty) = P_{(\theta,\sigma)}(\tau_- < \infty)P_{\theta_0}(\tau_+ = \infty) + P(\tau_{-M} < \infty, M > 0).$$

As $\theta_0, \theta \to 0$, from Lemma 6 of Wu (1999), we have

$$P_{(\theta,\sigma)}(\tau_- < \infty) = 1 - \sqrt{2}\frac{\theta}{\sigma}e^{-\frac{\theta}{\sigma}\rho}(1 + o(\theta^2)),$$

$$P_{\theta_0}(\tau_+ = \infty) = -\sqrt{2}\theta_0 e^{\theta_0\rho}(1 + o(\theta_0^2)).$$

Next, by conditioning on the value of M for $M > 0$, and noting that

$$P_{\theta_0}(M < x) = 1 - P_{\theta_0}(\tau_x < \infty),$$

and as $x \to \infty$ and θ_0 is small

$$P_{\theta_0}(\tau_x < \infty) \approx e^{2\theta_0(x+\rho)},$$

we can write

$$P(\tau_{-M} < \infty, M > 0) = -\int_0^\infty P_{(\theta,\sigma)}(\tau_{-x} < \infty)dP_{\theta_0}(\tau_x < \infty)$$

$$= -\int_0^\infty P_{\theta/\sigma}(\tau_{-x/\sigma} < \infty)dP_{\theta_0}(\tau_x < \infty)$$

$$= -\int_0^\infty e^{-2\frac{\theta}{\sigma}(\frac{x}{\sigma}+\rho)}de^{2\theta_0(x+\rho)} - \int_0^\infty \left(P_{\theta/\sigma}(\tau_{-x/\sigma} < \infty) - e^{-2\frac{\theta}{\sigma}(\frac{x}{\sigma}+\rho)}\right)$$

$$\times de^{2\theta_0(x+\rho)}$$

$$- \int_0^\infty e^{-2\frac{\theta}{\sigma}(\frac{x}{\sigma}+\rho)}d\left(P_{\theta_0}(\tau_x < \infty) - e^{2\theta_0(x+\rho)}\right)$$

$$- \int_0^\infty \left(P_{\theta/\sigma}(\tau_{-x/\sigma} < \infty) - e^{-2\frac{\theta}{\sigma}(\frac{x}{\sigma}+\rho)}\right)d\left(P_{\theta_0}(\tau_x < \infty) - e^{2\theta_0(x+\rho)}\right), \quad (6.6)$$

where in the last step, we split the integration into four terms. The first term of (6.6) on the right-hand side is equal to

$$-\frac{2\theta_0}{2(\theta/\sigma^2 - \theta_0)}e^{-2\theta\rho/\sigma + 2\theta_0\rho}.$$

For the second, third, and fourth terms of (6.6), we use the approximation

$$P_{\theta/\sigma}(\tau_{-x/\sigma} < \infty) - e^{-2\frac{\theta}{\sigma}(\frac{x}{\sigma}+\rho)} = E_{\theta/\sigma}e^{-2\frac{\theta}{\sigma}(\frac{x}{\sigma}+R_{x/\sigma})} - e^{-2\frac{\theta}{\sigma}(\frac{x}{\sigma}+\rho)}$$

$$= -2\frac{\theta}{\sigma}E_0(R_{x/\sigma} - \rho) + o(\theta),$$

and

$$P_{\theta_0}(\tau_x < \infty) - e^{2\theta_0(x+\rho)} = 2\theta_0(E_0 R_x - \rho)(1 + o(1)).$$

Thus, the second term of (6.6) is

$$4\frac{\theta\theta_0}{\sigma} \int_0^\infty (E_0 R_{x/\sigma} - \rho)dx;$$

the third term of (6.6) is

$$\left(P_{\theta_0}(\tau_+ < \infty) - e^{2\theta_0\rho}\right) e^{-2\theta_0\rho/\sigma} - 4\frac{\theta_0\theta}{\sigma} \int_0^\infty (E_0 R_x - \rho)dx,$$

and the fourth term of (6.6) is

$$\frac{4\theta_0\theta}{\sigma} \int_0^\infty (E_0 R_{x/\sigma} - \rho)d(E_0 R_x - \rho).$$

The lemma is proved by some simplifications.

6.6.2 Proof of Lemma 6.4

We first write

$$\begin{aligned}
E[\eta_M; \tau_{-M} = \infty] &= E_{\theta_0}\eta_M - E[\eta_M; \tau_{-M} < \infty] \\
&= E_{\theta_0}\eta_M - E_{\theta_0}\left[\eta_M e^{-2\frac{\theta}{\sigma}(\frac{M}{\sigma}+\rho)}\right](1 + o(\theta^2)),
\end{aligned}$$

as shown in Chapter 2.

From Lemma 6.1, we have

$$\begin{aligned}
E_{\theta_0}\eta_M &= \frac{E_{\theta_0}[\tau_+; \tau_+ < \infty]}{P_{\theta_0}(\tau_+ = \infty)} \\
&= \frac{1}{2\theta_0^2} - \frac{1}{4} + o(\theta_0).
\end{aligned}$$

On the other hand, we note that M is equivalent in distribution to $S_{\tau_+^{(K)}}$ and η_M is equivalent to $\tau_+^{(K)}$ in distribution, where

$$\tau_+^{(k)} = \inf\{n > \tau_+^{(k-1)} : S_n > S_{\tau_+^{(k-1))}}\}$$

is the kth ladder epoch and

$$K = \sup\{k > 0 : \tau_+^{(k)} < \infty\},$$

which is a geometric random variable with terminating probability $P(\tau_+ = \infty)$.

Thus, by conditioning on the value of K, we can obtain

$$E_{\theta_0}\left[\eta_M e^{-2\theta M/\sigma^2}\right] = P_{\theta_0}(\tau_+ = \infty)\frac{E_{\theta_0}\left[\tau_+ e^{-2\frac{\theta}{\sigma^2}S_{\tau_+}}; \tau_+ < \infty\right]}{\left(1 - E_{\theta_0}\left[e^{-2\frac{\theta}{\sigma^2}S_{\tau_+}}; \tau_+ < \infty\right]\right)^2}.$$

The lemma is proved by using the following two approximations, which are given by Lemmas 2.5 and 2.6 in Wu (2004):

$$1 - E_{\theta_0}\left[e^{-2\frac{\theta}{\sigma^2}S_{\tau_+}}; \tau_+ < \infty\right] = \sqrt{2}\left(\frac{\theta}{\sigma^2} - \theta_0\right)e^{-(2\frac{\theta}{\sigma^2}-\theta_0)\rho + (\frac{\theta}{\sigma^2}-\theta_0)\frac{\theta}{\sigma^2}}$$
$$\times(1 + o(\theta^2)),$$
$$E_{\theta_0}\left[\tau_+ e^{-2\frac{\theta}{\sigma^2}S_{\tau_+}}; \tau_+ < \infty\right] = -\frac{1}{\sqrt{2}\theta_0}e^{-(2\frac{\theta}{\sigma^2}-\theta_0)\rho + ((\frac{\theta}{\sigma^2})^2-\theta_0^2)/2}(1 + o(\theta^2)).$$

6.7 Case Study

6.7.1 IBM Stock Prices:

We return to the IBM stock price data in Section 2.4. Although the change of variance is much more obvious than the change in mean and has been analyzed in Chapter 2 and by several authors, here as a demonstration for our method, we suppose our initial goal is to detect a change in the yield rate by assuming a possible change in volatility.

As the data are already given offline, we make some manipulations to satisfy the assumptions for our model. We first find the standard deviation for the first 225 data as 0.009733 and then divide all the data by 0.009733, which makes the pre-change variance roughly 1.00. Then we subtract from all observations the estimated pre-change mean 0.0586 and switch the sign of all observations, which makes the pre-change mean roughly zero and post-change mean roughly 0.333. Next, we select $\theta_0 = -0.165$, which is roughly half of -0.333 by subtracting all observations by -0.165.

Next, we form the CUSUM process; the values are reported as follows and only two decimals are presented.

```
[1]   0.00 0.57 1.59 0.00 0.00 0.56 0.00 0.00 0.00 0.52
[11]  0.00 0.00 0.00 0.31 0.41 1.55 1.65 3.88 1.86 0.92
[21]  1.65 2.60 3.14 3.25 2.93 3.25 2.71 3.45 2.50 3.04
[31]  2.93 2.40 2.08 2.40 2.51 2.62 2.51 2.84 2.51 2.41
[41]  2.30 4.17 3.84 3.51 2.53 2.42 3.19 1.77 0.00 0.00
[51]  0.00 0.10 0.00 0.74 0.00 0.00 0.00 0.00 0.10 0.00
[61]  0.51 1.03 0.71 0.00 0.00 0.00 0.29 0.00 0.70 0.79
[71]  0.00 0.00 0.00 0.00 0.00 0.00 0.64 0.00 0.64 1.67
[81]  0.81 0.00 0.65 0.00 0.27 0.92 1.38 1.46 2.70 5.53
[91]  3.46 0.85 0.36 1.02 0.34 0.23 0.00 0.00 0.00 0.26
[101] 0.16 0.00 0.00 0.00 0.25 0.00 0.00 0.00 0.00 0.00
[111] 0.00 0.00 0.58 0.99 2.79 2.34 2.58 3.18 2.72 1.04
[121] 0.93 0.14 0.03 0.10 0.00 0.00 0.41 0.30 0.72 1.31
[131] 2.25 3.21 2.75 0.71 1.30 2.08 2.15 1.16 1.58 1.47
[141] 2.08 1.97 1.69 1.22 0.06 0.48 0.73 1.51 2.30 1.48
[151] 1.37 1.62 0.80 0.00 0.00 0.78 1.92 0.92 1.89 3.78
```

Table 6.2: Post-Change Mean and Variance Estimator for IBM Stock

d	$N - \hat{\nu}$	$\hat{\theta}$	$\hat{\sigma}$
10	18	0.594	0.815
15	19	0.914	0.851
20	44	0.458	1.877
25	46	0.632	1.917

[161] 5.71 4.30 4.01 2.80 2.52 1.86 2.85 3.11 4.11 4.01
[171] 3.90 4.92 4.25 5.27 5.73 7.35 8.60 5.22 2.68 2.02
[181] 3.21 3.10 2.81 2.52 1.67 1.57 3.13 3.21 3.48 3.38
[191] 4.41 4.30 4.96 4.47 4.75 4.45 4.15 2.90 1.86 2.12
[201] 1.83 1.54 1.24 0.39 0.28 0.36 0.44 0.52 0.00 0.00
[211] 0.45 0.35 1.36 0.88 0.77 2.56 2.08 2.16 3.01 4.46
[221] 6.73 7.23 4.93 4.82 4.71 4.21 5.49 6.39 4.88 4.97
[231] 4.27 4.36 4.26 4.15 4.24 5.13 7.87 10.69 17.37 15.47
[241] 12.94 10.90 9.94 12.19 15.60 17.30 16.74 18.24 16.08 16.65
[251] 15.41 17.12 16.79 20.16 24.89 29.08 28.72 38.90 32.71 29.96
[261] 31.96 38.76 37.51 37.13 35.62 34.41 38.51 45.51 49.15 54.60
[271] 45.81 46.63 44.67 47.04 50.77 49.37 45.48 49.16 44.96 41.51
[281] 42.91 39.52 39.12 39.31 40.68 39.10 36.38 31.80 30.60 31.31
[291] 29.58 32.75 34.04 33.37 31.88 31.49 34.18 31.83 32.28 30.25
[301] 27.18 27.07 29.93 27.39 27.28 28.52 30.32 27.77 29.01 29.18
[311] 28.26 26.27 25.90 26.06 25.16 23.74 23.89 19.69 19.08 19.47
[321] 19.62 23.36 23.78 24.46 22.26 24.52 25.48 24.04 25.53 24.89
[331] 25.32 24.95 24.84 23.40 21.46 22.13 23.61 24.31 25.82 29.32
[341] 27.81 13.68 33.03 32.05 35.79 32.71 32.90 30.18 29.50 29.39
[351] 27.59 29.47 30.22 30.40 26.88 29.61 28.36 29.98 31.92 33.30
[361] 37.16 33.99 34.19 36.53 32.17 30.00 31.66 29.79 28.23

For any value of d (at least 10), the last zero point of the CUSUM process is at the 210th observation, which is the estimated change-point. This shows that the method is very robust to the choice of d in terms of the change-point estimator. However, if we take the threshold as 5, there is a signal at observation 174, which can be thought of as a false signal.

Depending on the value of d, the sample size for estimating the post-change mean will change dramatically due to the large post-change variance. (Indeed, in this example, it is better to set the primary goal as detecting a change in variance with a possible change in mean.) The following table gives the summary for the sample size and estimators for post-change mean and variance for several values of d.

We see that due to large post-change variance estimators, the post-change mean estimator significantly overestimates the true value, as we expected from

Corollary 6.1.

6.7.2 Nile River Data

Here, we reconsider the Nile River data from a different point of view. As we noted from Chapters 2 and 4, the pre-change standard deviation is slightly higher than the post-change standard deviation. So we can make a finer analysis by assuming that there is a possible change in variance.

In this case, we shall further divide the standardized data by 1.08, which makes the pre-change standard deviation 1.00 and post-change standard deviation $\sigma = 0.926$.

As an effect of this transformation, the pre-change mean becomes -0.909 and the post-change mean becomes 0.926, which shows a slight discrepancy.

The modified CUSUM process is calculated as follows:

```
 [1] 0.00 0.00 0.09 0.00 0.00 0.00 1.20 0.00 0.00 0.00
[11] 0.00 0.30 0.00 0.00 0.00 0.11 0.00 1.30 1.43 0.21
[21] 0.00 0.00 0.00 0.00 0.00 0.00 0.00 0.00 1.49 2.49
[31] 3.24 5.32 5.58 6.63 8.66 9.10 11.19 10.86 10.30 10.35
[41] 11.42 13.26 17.10 18.22 20.24 19.17 18.24 19.30 20.87 22.01
[51] 23.54 24.50 25.33 26.16 28.22 29.18 30.89 32.22 31.73 33.33
[61] 34.77 35.59 36.55 36.78 36.71 37.29 38.42 38.16 39.67 41.89
[71] 44.30 45.26 46.47 48.19 49.48 49.00 49.85 50.60 51.54 52.17
[81] 53.88 55.56 56.57 56.02 56.44 56.36 57.67 58.06 58.06 59.24
[91] 58.91 59.42 59.97 58.53 58.99 60.69 61.10 63.01 64.94 66.68
```

We see that the change-point estimator is still at $\hat{\nu} = 28$. The following table reports the post-change sample size, post-change mean estimator, and post-change standard deviation estimator.

We see that the bias of $\hat{\theta}$ is slightly smaller than the one in the equal variance case.

Listed also are the theoretical bias for $\hat{\theta}$ calculated from Corollary 6.1 with $\theta_0 = -0.909$, $\theta = 0.926 = \sigma$, which gives the theoretical bias as

$$E^{\nu}[\hat{\theta} - \theta | N > \nu] \approx \frac{1.496}{d}.$$

Although our primary goal is to study the quantitative behavior for the biases of post-change mean and post-change variance, a practical method to correct the biases can be developed by estimating the bias. One method is given in Whitehead(1986). By using the theoretical formula for the biases of $\hat{\nu}$ and $\hat{\theta}$, we can obtain the bias-corrected estimator $\tilde{\nu}$ and $\tilde{\theta}$ by solving the following three equations simultaneously:

$$\hat{\nu} = \tilde{\nu} + \text{bias of } \hat{\nu}(\tilde{\theta}, \tilde{\sigma}),$$

$$\hat{\theta} = \tilde{\theta} + \frac{a(\tilde{\theta}, \tilde{\sigma})}{\sqrt{d}},$$

Table 6.3: Post-Change Mean and S.D. Estimator in Nile River Data

d	size	post-change mean	bias	theo. approx.	post-change s.d.
5	4	1.330	0.404	0.299	0.588
10	9	1.244	0.318	0.150	0.713
15	15	1.140	0.214	0.100	1.141
20	17	1.191	0.265	0.075	1.089
25	25	1.013	0.087	0.060	1.085
30	29	1.065	0.139	0.050	1.031
35	34	1.047	0.121	0.043	0.996
40	42	0.997	0.071	0.037	0.963
45	44	1.029	0.103	0.033	0.964
50	50	1.012	0.086	0.030	0.936
55	54	1.029	0.103	0.027	0.911
60	68	0.893	-0.033	0.025	0.928
65	72	0.926	0.000	0.023	0.924

$$\hat{\sigma}^2 \;=\; \tilde{\sigma}^2 + \text{bias of } \hat{\sigma}^2(\tilde{\theta}, \tilde{\sigma}).$$

There is no explicit solution for the corrected estimators due to the complicated form for the biases. A slightly less satisfactory, but convenient solution is to replace $\tilde{\theta}, \tilde{\sigma}$ in the biases by $\hat{\nu}$ and $\hat{\theta}$ as given in (6.1) and (6.4). We hope to present more practical methods and examples in a future communication.

7

Sequential Classification and Segmentation

7.1 Introduction

We start with a hidden Markov chain model. Let the parameter (state) process $\theta(t)$ be a two-state homogeneous Markov chain with states θ_0 and θ_1 and transition probabilities

$$P(\theta(t) = \theta_0|\theta(t-1) = \theta_1) = q \quad \text{and} \quad P(\theta(t) = \theta_1|\theta(t-1) = \theta_0) = p.$$

Given $\theta(t)$, let $\{X_t\}$ be a conditionally independent sequence of random variables following the exponential family distribution

$$dF_\theta(x) = \exp(\theta x - c(\theta))dF_0(x),$$

for $|\theta| < K(> 0)$, where $c(0) = c'(0) = 0$, $c''(0) = 1$, $\gamma = c^{(3)}(0)$, and $\kappa = c^{(4)}(0)$.

For the convenience of discussion, we assume $\theta_0 < 0 < \theta_1$ are a conjugate pair such that $c(\theta_0) = c(\theta_1)$ and $|\theta_i| < K$, for $i = 0, 1$.

For $\pi_0 = \pi$, we denote

$$\pi_n = P(\theta(n) = \theta_1|X_1, ..., X_n),$$

as the posterior probability of $\theta(n) = \theta_1$ given $X_1, ..., X_n$. Define

$$Y_n = \pi_n/(1 - \pi_n) \quad \text{and} \quad Z_n = \ln Y_n.$$

103

Then it can be shown that

$$Y_n = \frac{p + \bar{q}Y_{n-1}}{\bar{p} + qY_{n-1}} e^{(\theta_1 - \theta_0)X_n}$$

$$= Y_{n-1}\frac{\bar{q} + p/Y_{n-1}}{\bar{p} + qY_{n-1}} e^{(\theta_1 - \theta_0)X_n},$$

where $\bar{p} = 1 - p$ and $\bar{q} = 1 - q$.

Thus, the online Bayesian classification rule (nonlinear filter) will be

$$\tilde{\theta}(n) = \theta_1 I_{[Y_n \geq 1]} + \theta_0 I_{[Y_n < 1]}$$

$$= \theta_1 I_{[Z_n \geq 0]} + \theta_0 I_{[Z_n < 0]},$$

where

$$Z_n = \ln Y_n$$

$$= Z_{n-1} + \ln(\bar{q} + pe^{-Z_{n-1}}) - \ln(\bar{p} + qe^{Z_{n-1}}) + (\theta_1 - \theta_0)X_n.$$

As $p, q \to 0$ at the same order, we see that

$$Z_n \approx Z_{n-1} + (\theta_1 - \theta_0)X_n,$$

for $-(\ln p)Z_n = o(1)$ and $(\ln q)Z_n = o(1)$; while

$$Z_n \approx -\ln(1/p) + (\theta_1 - \theta_0)X_n,$$

for $-\ln(1/p)Z_n \to \infty$, and

$$Z_n \approx \ln(1/q) + (\theta_1 - \theta_0)X_n$$

for $\ln(1/q)Z_n \to \infty$. For this reason, Khasminskii, Lazareva, and Stapleton (1994) considered an asymptotic online filter by using

$$\tilde{Z}_n = (\theta_1 - \theta_0)X_n - \ln\left(\frac{1}{p}\right) I_{[\tilde{Z}_{n-1} \leq -\ln(\frac{1}{p})]}$$

$$+ \ln\left(\frac{1}{q}\right) I_{[\tilde{Z}_{n-1} \geq \ln(\frac{1}{q})]} + \tilde{Z}_{n-1} I_{[-\ln(\frac{1}{p}) < \tilde{Z}_{n-1} < \ln(\frac{1}{q})]},$$

as an approximation to Z_n. The version of continuous-time analog is given in Khasminskii and Lazareva (1992).

In this chapter, we consider this approximation in a more convenient form by using the dam process[Prabhu (1965)] as the classification process. Define

$$D_n = \min(b, \max(0, D_{n-1} + X_n)), \quad \text{for} \quad D_0 = d, \tag{7.1}$$

where b is the capacity level. For a classification limit a $(0 < a < b)$, we define the classifier as

$$\hat{\theta}(n) = \theta_1 I_{[D_n \geq a]} + \theta_0 I_{[D_n < a]}. \tag{7.2}$$

The dam process is a natural generalization of the CUSUM process, which is used for detecting a single change point. For this randomly changing parameter model, Page (1954) considered sequential sentence and correction after detection in a production system when defective items are produced occasionally. Beattie (1962) and Wasserman and Wadworth (1989) formally considered a double CUSUM process. A description of the double CUSUM procedure is given in the last section. From a system monitoring point of view, Hines (1974) considered using an EWMA process as a classification process. Recent applications can be seen in Avery and Henderson (1996), Fu and Curnow (1994), Halpern (2000) for locating change segments and heterogeneous segments in DNA sequence analysis and Assaf (1997), Khasminskii and Lazareva (1996), Khasminskii, Lazareva and Stapleton (1994) and Yao (1984) from a signal processing point of view.

In this Chapter, we consider the online classification based on the dam process under a typical nonparametric model in order to demonstrate the procedure. We assume that averages of the consecutive sojourn time on state θ_0 and θ_1 have a finite limit. In Section 7.2, we first consider the optimal design of a and b. The asymptotic limiting average error rate is derived as the average sojourn times on θ_0 and θ_1 are large. Simple formulas for the optimal a and b are obtained.

We further propose an offline segmentation procedure by modifying the online classification procedure. In Chapter 2, the estimation of a single change-point after detected by a CUSUM procedure has been considered. Here, we extend the idea to sequential segmentation by locating the change points (switching points) one by one as the last time point the process D_n is left 0 or b before it reaches b or 0, respectively. After a switching point is located, say as the end of a segment of θ_0 (or θ_1), a dam process is restarted from this located change point with initial value b (or 0). The memory requirement is quite small and computing intensity is light compared with window-limited search and Bayesian methods. The limiting average error rate under the online optimal design is derived.

In Section 7.4, we consider the second-order approximation for the error rate as both θ_0 and θ_1 approach 0 at the same order. Numerical and simulation results in the normal case are used for illustration, which shows that the approximations are quite satisfactory.

In Section 7.5, we briefly discuss the double CUSUM procedure and show its difference with the proposed procedure. A direct sequential segmentation for a multistate parameter process is proposed by generalizing the CUSUM procedure.

7.2 Online Classification

In this section, in order to consider the optimal design of a and b and compare it with other methods, we assume the following ergodicity condition.

(A1) The average consecutive sojourn times of $\theta(t)$ on θ_0 and θ_1 have constant finite limits T_0 and T_1, respectively.

Thus, in the infinite time horizon case, we can define the limiting average first type and second type of error rate as

$$\alpha = \lim_{n\to\infty} \frac{1}{n} \sum_{i=1}^{n} I_{[\hat{\theta}(i)=\theta_1;\theta(i)=\theta_0]}$$

and

$$\beta = \lim_{n\to\infty} \frac{1}{n} \sum_{i=1}^{n} I_{[\hat{\theta}(i)=\theta_0;\theta(i)=\theta_1]},$$

respectively. Thus, $\alpha+\beta$ will be the long-run average total error rate and can be calculated as the ratio between the total average length of the first and second types of error in a cycle with the cycle length $T_0 + T_1$.

To evaluate the error rate, we further assume that

(A2) Both T_0 and T_1 approach ∞, and T_1 goes to ∞ at a slower order than T_0 but no slower than $(\ln T_0)^{1+\epsilon}$ for some $\epsilon > 0$.

Then we shall be able to derive the asymptotic approximations for the error rate in a cycle at the first-order.

Obviously, as T_0 and T_1 approach ∞, a and b approach ∞ as well; otherwise the error rate will approach a constant.

We first consider the total length for the second type of error, which consists of two parts.

(1) The delay response time from the time when $\theta(t)$ is changed from θ_0 to θ_1 until D_n reaches a.

(2) The random errors caused by D_n down-crossing a during the sojourn period T_1 on θ_1.

For the first part, the average delay response time is asymptotically equivalent to the average delay detection time for detecting a change from θ_0 to θ_1 by the CUSUM procedure with control limit a.

The following notations will be used in our presentation. Denote

$$S_n = S_0 + \sum_{i=1}^{n} X_i$$

for $n \geq 0$. Denote by $P_{\theta_i}(.)$ the measure under the parameter θ_i for $i = 0, 1$ and $P_{\theta_i\theta_j}(.)$ the joint measure when both θ_i and θ_j appear for $i \neq j$ for $i, j = 0, 1$. For $S_0 = 0$, let S_n' be an independent copy of S_n and

$$M = \sup_{n\geq 0} S_n' \quad \text{and} \quad m = \inf_{n\geq 0} S_n'.$$

Also, for $S_0 = 0$, we denote

$$\begin{aligned}
\tau_x &= \inf\{n > 0 : S_n > x\}, \quad \text{for } x > 0, \\
&= \inf\{n > 0 : S_n \leq x\}, \quad \text{for } x \leq 0,
\end{aligned}$$

and $\tau_- = \tau_0$ and τ_+ as the limit of τ_x as $0 < x \to 0$.

In addition, as $x \to \infty$, we denote

$$R_\infty = \lim_{x\to\infty} (S_{\tau_x} - x), \quad \text{and} \quad R_{-\infty} = \lim_{x\to\infty} (S_{\tau_{-x}} + x)$$

as the overshoot and

$$\rho_+(\theta_1) = E_{\theta_1} R_\infty \quad \text{and} \quad \rho_-(\theta_0) = E_{\theta_0} R_{-\infty}.$$

The following result is proved in an unpublished technical report by Pollak and Siegmund (1986). For the sake of convenience, we give a different form and a modified proof in the appendix.

Lemma 7.1: *As $a \to \infty$ and $T_0 \to \infty$, the average delay response time for D_n to reach a after a switch from θ_0 to θ_1 is*

$$\frac{1}{c'(\theta_1)}(a + K_1 + o(1)),$$

where

$$K_1 = \rho_+(\theta_1) - E_{\theta_0}(M) + E_{\theta_1}(m) P_{\theta_0\theta_1}(\tau_{-M} < \infty) + E_{\theta_0\theta_1}[S_{\tau_{-M}} + M; \tau_{-M} < \infty],$$

where $E[X; A] = E[X I_A]$.

The next lemma gives the total expected length for the random second type of error.

Lemma 7.2: *The total expected length of the second type of random error is*

$$T_1 e^{-(\theta_1 - \theta_0)(b-a)} E_{\theta_0} e^{(\theta_1 - \theta_0) R_{-\infty}} (1 + o(1)).$$

Proof: As $a, b - a \to \infty$, from Wald's likelihood ratio identity, due to the exponential rate of convergence of D_n, we have

$$
\begin{aligned}
\lim_{n \to \infty} P_{\theta_1}(D_n < a) &= P_{\theta_1}(m < -(b-a))(1 + o(1)) \\
&= P_{\theta_1}(\tau_{-(b-a)} < \infty)(1 + o(1)) \\
&= E_{\theta_0} e^{(\theta_1 - \theta_0) S_{\tau_{-(b-a)}}}(1 + o(1)) \\
&= e^{-(\theta_1 - \theta_0)(b-a)} E_{\theta_0} e^{(\theta_1 - \theta_0) R_{-\infty}}(1 + o(1)).
\end{aligned}
$$

Note that the average delay response time is at the lower order of T_1, and we get the expected result.

Thus, the total average length of the second type of error in a cycle is

$$\frac{1}{c'(\theta_1)}(a + K_1 + o(1)) + T_1 e^{-(\theta_1 - \theta_0)(b-a)} E_{\theta_0} e^{(\theta_1 - \theta_0) R_{-\infty}}(1 + o(1)). \tag{7.3}$$

Similarly, the total average length for the first type of error is

$$\frac{1}{-c'(\theta_0)}(b - a + K_2 + o(1)) + T_0 e^{-(\theta_1 - \theta_0)a} E_{\theta_1} e^{-(\theta_1 - \theta_0) R_\infty}(1 + o(1)), \tag{7.4}$$

where

$$K_2 = -\rho_-(\theta_0) + E_{\theta_1}(m) - E_{\theta_0}(M) P_{\theta_1\theta_0}(\tau_{-m} < \infty) + E_{\theta_1\theta_0}[S_{\tau_{-m}} + m; \tau_m < \infty].$$

The total error length is the sum of (7.3) and (7.4). By considering a and $b - a$ as two variables, we minimize the total expected length of error. The results can be summarized in the following theorem.

Theorem 7.1: *Under Conditions (A1) and (A2), the optimum a and $b - a$ are given by*

$$a = \frac{1}{\theta_1 - \theta_0}(\ln[(\theta_1 - \theta_0)c'(\theta_1)T_0 E_{\theta_1} e^{-(\theta_1 - \theta_0)R_\infty}] + o(1)),$$

$$b - a = \frac{1}{\theta_1 - \theta_0}(\ln[-(\theta_1 - \theta_0)c'(\theta_0)T_1 E_{\theta_0} e^{(\theta_1 - \theta_0)R_{-\infty}}] + o(1)),$$

and under this optimal design the minimum error rate in a cycle is

$$\alpha + \beta = \frac{1}{T_0 + T_1}\{\frac{1}{(\theta_1 - \theta_0)c'(\theta_1)} \ln\left[(\theta_1 - \theta_0)c'(\theta_1)T_0 E_{\theta_1} e^{1-(\theta_1 - \theta_0)R_\infty}\right]$$
$$- \frac{1}{(\theta_1 - \theta_0)c'(\theta_0)} \ln\left[-(\theta_1 - \theta_0)c'(\theta_0)T_1 E_{\theta_0} e^{1+(\theta_1 - \theta_0)R_{-\infty}}\right]$$
$$+ \frac{K_1}{c'(\theta_1)} - \frac{K_2}{c'(\theta_0)} + o(1)\}.$$

Remark 7.1: It can be seen that for online classification, under the optimum design, the delay response time is the major factor for the error rate.

Remark 7.2: T_0 and T_1 can be estimated adaptively in the finite-time horizon from an initial guess or estimation. Also, from Theorem 7.1, we can see that the error rate is not sensitive to the choice of T_0 and T_1 as only the log-scales of T_0 and T_1 are involved.

Remark 7.3: Theorem 7.1 gives the second-order result, which is consistent with the result of Khasminskii, Lazareva, and Stapleton (1994) where only the first-order result is conjectured as in Theorem 2.1.

7.3   Offline Segmentation

The online classification is used when an immediate classification has to be made. That means no lag time is allowed. For offline segmentation in a large data set, we can carry out a correction and smoothing following online classification in order to locate the change (switching) points.

Under the hidden Markov chain model, one can use the nonlinear filtering and interpolation techniques to get an estimator for $\theta(t)$ which are Bayes or ML estimators conditional on the whole data set; see Viterbi (1969), Yao (1984) and Churchill (1989). For applications to signal processing, we refer to Elliott, Aggoun and Moore (1995) for finite-state hidden Markov chains.

Chapter 2 considers the estimation of change-point after a change is detected by the CUSUM procedure in the exponential family. The change-point is estimated as the last zero point of the CUSUM process before the detection time. Under our model, the change-point is indeed detected one by one. So we can locate the change-points one by one by keep tracking the last time point D_n left 0 (or b) before it reaches b (or 0). After a change- (switching) point is located, the dam process is restarted from the estimated last change-point at b (or 0) until another change-point is detected by reaching 0 (or b), respectively. The procedure can be described by the following algorithm:

Suppose $\theta(0) = \theta_0$. *Let* $D_0 = 0$ *and set*

$$T_0^{(0)} = S_0^{(1)} = 0.$$

For $k \geq 0$, *we define* $S_k^{(0)}, T_k^{(0)}, S_k^{(1)}$, *and* $T_k^{(1)}$ *recursively as follows:*

$$
\begin{aligned}
T_{k+1}^{(1)} &= \inf\{n > S_k^{(1)} : \ D_n^{(0)} = b\}, \\
S_{k+1}^{(0)} &= \max\{S_k^{(0)} \leq n < T_{k+1}^{(1)} : \ D_n^{(0)} = 0\}, \\
T_{k+1}^{(0)} &= \min\{n > S_{k+1}^{(0)} : \ D_n^{(1)} = 0\}, \\
S_{k+1}^{(1)} &= \max\{S_{k+1}^{(0)} \leq n < T_{k+1}^{(0)} : \ D_n^{(1)} = b\},
\end{aligned}
$$

where we denote $D_n^{(i)}$ *as the restarted dam process after each* $S_{k+1}^{(i)}$ *without changing the time index.*

The corrected and smoothed estimation for $\theta(t)$ *after offline segmentation is defined as*

$$\theta^*(n) = \theta_0 I_{[S_k^{(1)} < n \leq S_{k+1}^{(0)}]} + \theta_1 I_{[S_{k+1}^{(0)} < n \leq S_{k+1}^{(1)}]},$$

for $k \geq 0$.

The correction and smoothing procedure mainly reduces the delay response time and also smoothes out all the random errors that cross the classification limit.

The total length of error after correction and smoothing consists of two different parts.

(1) The absolute bias of $S_k^{(0)}$ and $S_k^{(1)}$ as the estimators for the change-point from θ_0 to θ_1 and from θ_1 to θ_0, respectively.

(2) The periods of false segments.

We still restrict the discussion to the asymptotic case and consider the total length of the first type of error in a cycle. For the second part, we note that a first type of false segment happens when D_n reaches b. In the stationary state,

$$P_{\theta_0}(D_\infty = b) = P_{\theta_0}\left(\sup_{n \geq 0} S_n > b\right) = O\left(e^{-b(\theta_1 - \theta_0)}\right).$$

Thus the total number of false segments of the first type is at the order of

$$O(T_0 e^{-b(\theta_1 - \theta_0)}) = O\left(\frac{1}{c'(\theta_1)(\theta_1 - \theta_0)} e^{-(b-a)(\theta_1 - \theta_0)}\right).$$

For each of these false segments of the first type, the expected total length is equal to $-b/c'(\theta_0)$ at the first order from Lemma 7.1. Thus, the total length of the first type of error due to false segments is at the order

$$O\left(-\frac{b}{c'(\theta_0)c'(\theta_1)(\theta_1-\theta_0)}e^{-(b-a)(\theta_1-\theta_0)}\right),$$

which is a negligible term $(o(1))$ under the optimal design and Condition (A2).

For the first part, Wu (2004a) considers the bias for the estimation of one change-point in the quasistationary state after a change is detected by a CUSUM process. Theorem 2.1 in Wu (2004a) can be used to find the absolute bias for $S_k^{(0)}$ and $S_k^{(1)}$. The result is summarized in the following theorem.

Theorem 7.2: *Under the optimal design in Theorem 7.1, the absolute bias of $S_k^{(0)}$ is*

$$B_0 \;=\; E_{\theta_0\theta_1}[\tau_{-M};\tau_{-M}<\infty] + P_{\theta_0\theta_1}(\tau_{-M}<\infty)\frac{E_{\theta_1}[\tau_-;\tau_-<\infty]}{P_{\theta_1}(\tau_-=\infty)}$$
$$+E_{\theta_0\theta_1}[\sigma_M;\tau_{-M}=\infty],$$

and the absolute bias of $S_k^{(1)}$ is

$$B_1 \;=\; E_{\theta_1\theta_0}[\tau_{-m};\tau_{-m}<\infty] + P_{\theta_0\theta_1}(\tau_{-m}<\infty)\frac{E_{\theta_0}[\tau_+;\tau_+<\infty]}{P_{\theta_0}(\tau_+=\infty)}$$
$$+E_{\theta_1\theta_0}[\gamma_m;\tau_{-m}=\infty],$$

where σ_M is the maximum point of S_n' and γ_m is the minimum point of S_n' under $P_{\theta_1}(.)$. The total long-run average error rate is asymptotically

$$\frac{1}{T_0+T_1}(B_0+B_1+o(1)).$$

7.4   Second-Order Approximations

In this section, in order to evaluate the online and offline error rates, we further assume the following locality condition:

(A3) θ_0 and θ_1 approach zero at the order of $o\left(\min\left(\frac{1}{a^{1-\epsilon}},\frac{1}{(b-a)^{1-\epsilon}}\right)\right)$ for some $\epsilon>0$.

Under this condition, if $F_0(x)$ is nonlattice, the strong renewal theory given in Section 1.4 says that asymptotic results are still valid.

In the online case, Pollak and Siegmund (1985) showed the following second-order approximation, and a simpler proof following the proof for Lemma 7.1 in the normal case is given in the appendix.

Lemma 7.3: *Under conditions (A1)–(A3), the average delay response time from θ_0 to θ_1 is*

$$\frac{1}{c'(\theta_1)}\left\{a + 2\rho_+ - \frac{3}{2(\theta_1 - \theta_0)} - \frac{1}{2}(\rho_+ + \rho_-) - \frac{1}{4}(\rho_+ + \rho_-)^2\right.$$

$$\left. + \theta_1 Var_0(R_\infty) + o(\theta_1)\right\},$$

where $\rho_\pm = \rho_\pm(0)$. In the normal case, it becomes

$$\frac{1}{\theta_1}\left[a + 2\rho_+ - \frac{3}{4\theta_1} + \frac{\theta_1}{4} + o(\theta_1)\right],$$

where $\rho_+ \approx 0.583$.

Similarly, we can write down the result for the delay response time from θ_1 to θ_0. In the normal case, we summarize the result in the online case as follows.

Corollary 7.2: *In the normal case, under Conditions (A1)–(A3), the optimal values for a and b are given by*

$$a = \frac{1}{2\theta_1}[\ln(2\theta_1^2 T_0) - 2\theta_1\rho_+] + o(1),$$

$$b - a = \frac{1}{2\theta_1}[\ln(2\theta_1^2 T_1) - 2\theta_1\rho_+] + o(1),$$

and under the optimal design, the online limit average error rate is

$$\alpha + \beta = \frac{1}{2\theta_1^2(T_0 + T_1)}[\ln(4\theta_1^4 T_0 T_1) - 1 + 4\rho_+\theta_1^2 + o(\theta_1^2)].$$

The proof is obtained by noting that in the normal case,

$$E_{\theta_1} e^{-(\theta_1 - \theta_0)R_\infty} = e^{-2\theta_1\rho_+} + o(\theta_1^3),$$

as noted in Siegmund (1985, Chap. 10).

In the offline case, Wu (2004) considers the bias of the change-point estimator after a change is detected by a CUSUM procedure in the exponent family case, and the following corollary gives the result in the normal case.

Corollary 7.3: *In the normal case, the absolute bias of the change-point estimator is*

$$\frac{3}{4\theta_1^2} - \frac{1}{8} + o(1),$$

and the asymptotic error rate is

$$\frac{1}{T_0 + T_1}\left(\frac{3}{2\theta_1^2} - \frac{1}{4} + o(1)\right).$$

Remark 7.4: As noted by Wu (2004), the absolute bias is slightly larger than the one in the fixed sample case as considered in Wu (1999) in the constant term, but the difference is almost ignorable.

As a numerical comparison, we conduct a simulation study. For $T_0 = T_1 = 100$ and total time $=100,000$, we take $\theta_1 = 0.25$ and 0.5; the corresponding designs for a given in Corollary 7.1 are 4.47 and 3.33, respectively.

For the online case, the Bayes classifier gives the average error rates as 0.191 and 0.076. The dam process gives the average error rate as 0.190 and 0.078, respectively; while the approximated values are 0.188 and 0.082 from Corollary 7.1, respectively.

For the offline case, we only consider the case of $\theta_1 = 0.5$ with total time 10,000 based on a simulated path. With $a = 3.33$ and $b = 6.66$, we use the proposed offline segmentation procedure to locate the change-points one by one. The located change-points are

> 0;91;200;300;402;502;600;700;792;912;994;1102;1200;1298;1396;
> 1500;1602;1696;1805;1900;2000;2124;2198;2300;2401;2498;2600;
> 2696;2799;2900;3000;3094;3199;3298;3401;3499;3599;3697;3797;
> 3897;3995;4101;4202;4304;4403;4505;4603;4702;4798;4899;5003;
> 5097;5200;5298;5403;5500;5603;5692;5821;5905;6002;6104;6199;
> 6293;6410;6495;6596;6700;6796;6904;6999;7102;7198;7306;7397;
> **7446;7454;**7502;7604;7699;7800;7902;7999;8101;8188;8303;
> 8398;8500;8602;9694;8800;8901;8998;9109;9200;9300;9400;9500;
> 9595;9701;9800;9899;10000;

In the above sequence of located segments, there is only one false segment (7446:7454] with length 8. The total sum of absolute bias is 299, which gives the total average error rate as 0.031. The theoretical approximation in Corollary 3.1 gives the value 0.029. We see that the approximations are quite accurate in both the online and offline cases.

7.5 Discussion and Generalization

7.5.1 Some Discussion

In this chapter, we propose a sequential segmentation procedure based on a generalization of the CUSUM process. Our main emphasis is to show that after proper correction and smoothing, we can locate the change-points one by one without increasing too much computation. Although the results are given for large average sojourn times, the technique can be used quite flexibly in many practical situations, for example, locating a single change-segment.

From a modeling point of view, the hidden Markov chain models have been discussed extensively, particularly for computational convenience. The method proposed here is less model-dependent and quite convenient, particularly from a segmentation point of view.

7.5.2 Double CUSUM Procedure

From a quality inspection point of view, Page (1954) considered a double CUSUM procedure; see also Beattie (1962) and Wasserman and Wadworth (1989). In the double CUSUM procedure, two limits, say A and B, are designed similar to a and b. Once a CUSUM procedure with reflecting boundary 0 (or B) up-crosses (down-crosses) A, a new CUSUM procedure starts from B (or 0).

Apparently, the double CUSUM procedure provides a tool for online classification. Under the same conditions, the first-order stationary error length in a cycle is about

$$T_0 e^{-(\theta_1-\theta_0)A} E_{\theta_1} e^{-(\theta_1-\theta_0)R_\infty} \frac{B-A}{-c'(\theta_0)} + \frac{A}{c'(\theta_1)}$$

$$+ T_1 e^{-(\theta_1-\theta_0)(B-A)} E_{\theta_0} e^{(\theta_1-\theta_0)R_{-\infty}} \frac{A}{c'(\theta_1)} + \frac{B-A}{-c'(\theta_0)},$$

where the first term is the total first-type error length caused by the false signals in cycle, and the second term is the delay response time for the change from θ_0 to θ_1. After some algebra calculations, we see that optimal A and B satisfy

$$A = \frac{1}{\theta_1-\theta_0} \ln[c'(\theta_1)(B-A)T_0 E_{\theta_1} e^{-(\theta_1-\theta_0)R_\infty}] + o(1),$$

$$B - A = \frac{1}{\theta_1-\theta_0} \ln[-c'(\theta_0)AT_1 E_{\theta_0} e^{(\theta_1-\theta_0)R_{-\infty}}] + o(1).$$

Under this optimal design, the stationary error rate can be derived at the first order as

$$\frac{1}{T_0+T_1}\{\frac{1}{(\theta_1-\theta_0)c'(\theta_1)}(\ln(T_0) + \ln\ln(T_0) + O(1))$$

$$+ \frac{1}{-(\theta_1-\theta_0)c'(\theta_0)}(\ln(T_1) + \ln\ln(T_1) + O(1))\},$$

and the details are deleted. Thus, the double CUSUM procedure is less efficient as the dam process at the second-order from an online classification point of view. However, for offline segmentation, we can similarly define a locating process, which gives the same asymptotic error rate as the proposed procedure.

7.5.3 Multistate Change Model

If we are only interested in offline segmentation, the proposed segmentation procedure indeed consists a sequence of CUSUM procedures that locates the change-points one by one. In fact, we can generalize the method to recursive segmentation when the parameter changes in several states, say $\{\theta_0, \theta_1, ..., \theta_K\}$, as follows.

Suppose the sojourn time on the initial state θ_k is T_k, and at the end of staying on θ_k, $\theta(t)$ can change to any one of the θ_j for $j \neq k$.

Suppose the initial parameter is θ_{j_0}. We can run K CUSUM procedures simultaneously by defining

$$\tau_j^{(j_0)} = \inf\{n > 0 : Y_n^{(j_0 j)} = \max(0, Y_{n-1}^{(j_0 j)} + (\theta_j - \theta_{j_0}) X_n - (c(\theta_j) - c(\theta_{j_0}))) > d_j^{(j_0)}\},$$

where $d_j^{(j_0)}$ are selected such that

$$E_{j_0} \tau_j^{(j_0)} = ARL_{j_0}$$

are the same for all $j \neq j_0$.

Denote

$$\tau_{j_1}^{(j_0)} = \min_{j \neq j_0} \tau_j^{j_0}$$

and

$$j_1 = \operatorname{argmin}_{j \neq j_0} \tau_j^{(j_0)}.$$

Then, we locate

$$S_1^{(j_0 j_1)} = \max\{n < \tau_{j_1}^{(j_0)} : \quad Y_n^{(j_0 j_1)} = 0\}$$

as the first change-point from θ_{j_0} to θ_{j_1}. Now, starting from $S_1^{(j_0 j_1)}$ and treating θ_{j_1} as the new starting parameter, we run another K CUSUM procedures until another change is detected. And the procedure continues. A sequence of change-point and changed states is located recursively as $S_k^{(j_{i-1} j_i)}$ and θ_{j_i} for $i \geq 1$.

The limiting average error rate can be studied under certain specific models, such as the hidden Markov model as considered in Khasminskii and Zeitouni (1996), Elliott, Aggoun, and Moore (1995) with applications to DNA sequence segmentation as considered in Churchill (1989), Boys, Henderson, and Wilkinson (2000) and Braun, Braun, and Mueller (2000).

7.6   Proofs

For $S_0 = x$, define

$$N_x = \inf\{n > 0 : \ S_n \leq 0; \ or \ > a\}.$$

Note that for the CUSUM procedure with boundary a, the average run length under $P_1(.)$ is

$$ARL_1 = E_{\theta_1}(N_0) / P_{\theta_1}(S_{N_0} > a)$$

from Chapter 1. In the quasistationary state case, when a change from θ_0 to θ_1 occurs, D_n is asymptotically distributed as M under $P_{\theta_0}(.)$. Thus, the average delay response time is

$$E_{\theta_0 \theta_1}(N_M) + \frac{E_{\theta_1}(N_0)}{P_{\theta_1}(S_{N_0} > a)} P_{\theta_0 \theta_1}(S_{N_M} \leq 0)$$

$$= \frac{1}{c'(\theta_1)} \left[E_{\theta_0\theta_1}(S_{N_M}) - E_{\theta_0}(M) + \frac{E_{\theta_1}(S_{N_0})}{P_{\theta_1}(S_{N_0} > a)} P_{\theta_0\theta_1}(S_{N_M} \leq 0) \right].$$

As $a \to \infty$,

$$P_{\theta_0\theta_1}(S_{N_M} \leq 0) \to P_{\theta_0\theta_1}(\tau_{-M} < \infty)$$

$$P_{\theta_1}(S_{N_0} > a) \to P_{\theta_1}(\tau_- = \infty).$$

Furthermore,

$$\frac{E_{\theta_1}(S_{N_0})}{P_{\theta_1}(S_{N_0} > a)} = \frac{E_{\theta_1}(S_{N_0}; S_{N_0} \leq 0)}{P_{\theta_1}(S_{N_0} > a)} + E_{\theta_1}(S_{N_0} | S_{N_0} > a)$$

$$= \frac{E_{\theta_1}[S_{\tau_-}; \tau_- < \infty]}{P_{\theta_1}(\tau_- = \infty)} + a + \rho_+(\theta_1) + o(1),$$

and

$$E_{\theta_0\theta_1}(S_{N_M}) = E_{\theta_0\theta_1}(S_{N_M}; S_{N_M} \leq 0) + E_{\theta_0\theta_1}(S_{N_M}; S_{N_M} > a)$$

$$= E_{\theta_0\theta_1}[S_{\tau_{-M}} + M; \tau_{-M} < \infty](1+o(1)) + (a+\rho_+(\theta_1))P_{\theta_0\theta_1}(\tau_{-M} = \infty)(1+o(1)).$$

Lemma 7.1 follows by combining the above approximations. A rigorous argument from strong renewal theorem along the lines of Chapter 1 shows that all the error terms are at the order $O(e^{-r(\theta_1-\theta_0)a})$ for some $r > 0$.

In the normal case, as $\theta_1 = -\theta_0 \to 0$ under the local conditions, we have

$$P_{\theta_0\theta_1}(\tau_{-M} < \infty) = \frac{1}{2}e^{-2\theta_1^2} + o(\theta_1^2),$$

from Wu (1999). Also from Siegmund (1979),

$$-E_{\theta_1}(m) = E_{\theta_0}(M) = \frac{1}{2\theta_1} - \rho_+ + \frac{\theta_1}{4} + o(\theta_1),$$

and

$$\rho_+(\theta_1) = \rho_+ + \frac{\theta_1}{4} + o(\theta_1).$$

On the other hand,

$$E_{\theta_0\theta_1}[S_{\tau_{-M}} + M; \tau_{-M} < \infty] = -\rho_+(\theta_1)P_{\theta_0\theta_1}(\tau_{-M} < \infty)(1+o(1))$$

$$= -\frac{1}{2}\left(\rho_+ + \frac{\theta_1}{4} + o(\theta_1)\right).$$

Lemma 7.4 follows by combining these approximations.

Similar techniques can be used to obtain Lemma 7.1 and Corollary 7.2 as in Wu (1999, 2004b). The details are omitted.

8
An Adaptive CUSUM Procedure

8.1 Definition

To deal with more general models for post-change means such as the linear post-change means, more powerful monitoring procedures need to be applied. Lai (1995) reviewed the state of art of detecting procedures and proposed a window-limited detecting procedure that generalized the generalized likelihood ratio detecting procedure considered by Siegmund and Venkatraman (1993). Yakir, Krieger, and Pollak (1999) and Krieger, Pollak, and Yakir (2003) specifically considered the linear post-change mean model.

In this chapter, we consider an adaptive CUSUM procedure that is quite flexible in detecting any parametric post-change means without loss of convenience of estimating the change-point and post-change parameters.

Note that the CUSUM procedure can be understood as a sequence of sequential tests with a two-sided boundary $[0, d)$. Whenever a test ends with crossing the boundary 0, a new test starts again until a test ends with crossing the boundary d. The change-point estimator is conveniently considered as the starting point of the last test, and the post-change mean estimator is the sample mean of the last test.

Naturally, the adaptive CUSUM procedure will be formed as a sequence of adaptive sequential tests [Robbins and Siegmund (1973, 1974)] with the post-change parameters being estimated adaptively for each test. To make our discussion concrete, we restrict the observations $\{X_1, ..., X_\nu, X_{\nu+1}, ...\}$ to an independent normal sequence of unit variance with pre-change mean 0 and post-change means $\mu_i(\theta)$ for $i > \nu$, where θ is the post-change parameter, which may be multidimensional.

Formally, based on the observations $\{X_{k+1}, ..., X_n\}$, we define

$$\theta_{k,n} = \theta(X_{k+1}, ..., X_n)$$

and

$$\theta_{k,k} = \theta_0 = \delta$$

for $k \geq 0$. For notational convenience, we write $\theta_n = \theta_{0,n}$.

Set $T_0 = 0$ and $\nu_0 = 0$. Define recursively, for $n > 0$,

$$T_n = \max\{0, T_{n-1} + \mu_n(\theta_{\nu_{n-1},n-1})(X_n - \frac{1}{2}\mu_n(\theta_{\nu_{n-1},n-1}))\},$$

and if $T_n > 0$, reset

$$\begin{aligned}
\nu_n &= \nu_{n-1}, \\
\theta_{\nu_n,n} &= \theta(X_{\nu_n+1}, ..., X_n),
\end{aligned}$$

and if $T_n = 0$, we update

$$\nu_n = n \quad \text{and} \quad \theta_{n,n} = \delta.$$

An alarm will be made at the time

$$N = \inf\{n > 0 : \ T_n > d\}.$$

After a change is detected, the change-point and post-change parameter will be estimated by

$$\hat{\nu} = \nu_N \quad \text{and} \quad \hat{\theta} = \theta_{\nu_N,N}. \tag{8.1}$$

8.2 Examples

8.2.1 Simple Change Model

In this case, let μ be the post-change mean, and let

$$\mu_{k,n} = \mu(X_{k+1}, ..., X_n)$$

be an estimator for μ based on observations $X_{k+1}, ..., X_n$. In particular, let

$$\bar{X}_{k,n} = \frac{1}{n-k}(X_{k+1} + ... + X_n).$$

Example 8.1: (Recursive mean estimator) For initial estimator δ and a tuning value $t \geq 0$, define

$$
\begin{aligned}
\mu_{k,n} = \mu_{k,n}(t) &= \frac{t}{t+n-k}\delta + \frac{n-k}{t+n-k}\bar{X}_{k,n} \\
&= \delta + \frac{n-k}{t+n-k}(\bar{X}_{k,n} - \delta) \\
&= \mu_{k,n-1} + \frac{1}{t+n-k}(X_n - \mu_{k,n-1}),
\end{aligned}
$$

with $\mu_{k,k} = \delta$. In particular,

$$
\begin{aligned}
\mu_n(t) = \mu_n &= \frac{t}{t+n}\delta + \frac{n}{t+n}\bar{X}_n \\
&= \mu_{n-1} + \frac{1}{t+n}(X_n - \mu_{n-1}),
\end{aligned}
$$

with $\mu_0 = \delta$.

Example 8.2: [Robbins and Siegmund (1974)]
Suppose δ is the smallest change magnitude to be detected. Then we define

$$
\mu_{k,n} = \delta + (\bar{X}_{k,n} - \delta)^+
$$

and

$$
\mu_n = \delta + (\bar{X}_n - \delta)^+,
$$

with $\mu_{k,k} = \mu_0 = \delta$.

8.2.2   Linear Post-Change Means

For $k \geq \nu$, suppose

$$
\mu_k = \beta(k - \nu)
$$

and define

$$
\beta_{\nu_n,n} = \sum_{k=\nu_n+1}^{n} (k - \nu_n)X_i / \sum_{k=\nu_n+1}^{n} (k - \nu_n)^2,
$$

as the least-square estimator. The CUSUM process becomes

$$
T_n = \max\{0, T_{n-1} + \beta_{\nu_{n-1},n-1}(n-1-\nu_{n-1})(X_n - \frac{1}{2}\beta_{\nu_{n-1},n-1}(n-1-\nu_{n-1}))\}.
$$

8.3   Simple Change Model

8.3.1   A Nonlinear Renewal Theorem

For given μ_0, we define

$$
S_n = S_0 + \sum_{i=1}^{n} \mu_{i-1}(X_i - \mu_{i-1}/2).
$$

For $S_0 = 0$ and $x > 0$, we define

$$\tau_x = \inf\{n > 0 : \ S_n > x\},$$

and for $x \leq 0$,

$$\tau_x = \inf\{n > 0 : \ S_n \leq x\}.$$

Denote by $H_k = \sigma(\mu_0, X_1, ..., X_k)$ the history up to time k. Due to the special feature of this chapter, we introduce some different notations from other chapters. Let $P_0(.)$ be the pre-change measure with mean 0, and $P_\mu(.)$ the post-change measure with mean μ.

We are interested in the probability $P_0(\tau_d < \infty | H_k)$ as $d \to \infty$.

Define a changed measure $P_0^*(.)$ such that

$$dP_0^*(.|H_n) = \exp(S_n)dP_0(.|H_n).$$

Under $P_0^*(.)$, given H_k, suppose $\mu_{k,n}$ almost surely converges to a random variable $\mu_{k,\infty}$.

Given $\mu_{k,\infty} = \mu$, suppose $X_1^*, ..., X_n^*, ...$ are conditional i.i.d. $N(\mu, 1)$ random variables and

$$S_n^* = \sum_{i=1}^{n} \mu(X_i^* - \mu/2).$$

Similarly, we define

$$\tau_d^* = \inf\{n > 0 : \ S_n^* > d\},$$

and $R_d^* = S_{\tau_d^*} - d$. From standard renewal theory, we see that for given $\mu \neq 0$, R_d^* converges weakly to R_∞^*, say. Define

$$\nu(\mu) = \lim_{d \to \infty} E_0^* e^{-R_d^*}.$$

Theorem 8.1 [Woodroofe (1990)]: *As $d \to \infty$, $(\mu_{k,\tau_d}, S_{\tau_d} - d)$ has the same limiting distribution as $(\mu_{k,\infty}, R_d^*)$.*

Thus, by using Wald' likelihood ratio identity by changing measure $P_0(.)$ to $P_0^*(.)$, we have

$$\begin{aligned}
\lim_{d \to \infty} P_0(\tau_d < \infty | H_k) &= \lim_{d \to \infty} E_0^*[e^{-S_{\tau_d}} | H_k] \\
&= e^{-d} E_0^*[\nu(\mu_{k,\infty})] \\
&= e^{-d - r_1(H_k)},
\end{aligned}$$

where we denote

$$e^{-r_1(H_k)} = E_0^*[\nu(\mu_{k,\infty})] \quad \text{and} \quad r_1 = r_1(H_0).$$

Example 8.1: (Recursive mean estimation)

For given H_k,

$$\mu_n = \mu_k + \sum_{j=k+1}^{n} \frac{1}{t+j}(X_j - \mu_{j-1}),$$

which almost surely converges to a normal random variable with mean μ_k and variance

$$\sigma_k^2 = \sum_{j=k+1}^{\infty} 1/(t+j)^2.$$

Thus, under $P_0^*(.)$,

$$
\begin{aligned}
e^{-r_1(H_k)} &= \lim_{d \to \infty} E_0^*[e^{-(S'_{\tau_d} - d)} | H_k] \\
&= \int \nu(y) \frac{1}{\sigma_k} \phi\left(\frac{y - \mu_k}{\sigma_k}\right) dy \\
&= \int \nu(\mu_k + \sigma_k y)\phi(y)dy.
\end{aligned}
\tag{8.2}
$$

In particular,

$$e^{-r_1} = \int \nu(\delta + \sigma_0 y)\phi(y)dy.$$

Example 8.2: [Robbins and Siegmund (1974)]
Under $P_0^*(.)$, let X_n follow $N(\mu_{n-1}, 1)$. Then,

$$
\begin{aligned}
X_1 &= \delta + Z_1, \\
X_2 &= \delta + (X_1 - \delta)^+ + Z_2 \\
&= \delta + Z_1^+ + Z_2 \\
&= \delta + W_1 + Z_2;
\end{aligned}
$$

and in general, we can write

$$
\begin{aligned}
X_n &= \delta + W_{n-1} + Z_n, \\
W_{n-1} &= \frac{1}{n-1}(W_1 + ... + W_{n-2} + Z_1 + ... + Z_{n-1})^+,
\end{aligned}
$$

where Z_i are iid N(0, 1) r.v.s.

It can be shown in the appendix that W_n strongly converges to a random variable with probability 1, say, W_∞. W_∞ has the same distribution as $(U(1) + B(1))^+$, where $B(t)$ is the standard Brownian motion and $U(t)$ is a continuous differentiable increasing process satisfying

$$\frac{dU(t)}{dt} = \frac{1}{t}(U(t) + B(t))^+, \quad \text{for} \quad 0 \le t \le 1.$$

It seems difficult to find an exact form for the distribution of W_∞.

8.3.2 Average Run Lengths

For given μ_0, we denote

$$S_n = S_0 + \sum_{i=1}^{n} \mu_{i-1}(X_i - \mu_{i-1}/2),$$

and define

$$N_x = \inf\{n > 0: \ S_n \le 0; \ \text{ or } \ > d\}, \quad \text{for} \ S_0 = x,$$

and N_0 as the two-sided boundary crossing time with $S_0 = 0$.

When there is no change, every time that T_n comes back to zero consists of a renewal point. Similar to the derivation of (2.57) in Siegmund (1985), we have

$$ARL_0 = E_0(N) = \frac{E_0(N_0)}{P_0(S_{N_0} > d)}. \tag{8.3}$$

To evaluate $E_0(N_0)$ and $P_0(S_{N_0} > d)$, we use a similar argument as in (8.61) and (8.69) of Siegmund (1985) and have

$$
\begin{aligned}
P_0(S_{N_0} > d) &= P_0(\tau_d < \infty) - P_0(\tau_d < \infty; S_{N_0} \le 0) \\
&= E_1[e^{-S_{\tau_d}}] - E_0[P_0(\tau_d < \infty | S_{N_0}); S_{N_0} \le 0] \\
&\approx e^{-d-r_1} - E_0[e^{-d-r_1(H_{N_0})+S_{N_0}}; S_{N_0} \le 0] \\
&= e^{-d}(e^{-r_1} - E_0^*[e^{-r_1(H_{N_0})}; S_{N_0} \le 0)] \\
&\approx e^{-d}(e^{r_1} - E_0^*[e^{-r_1(H_{\tau_-})}; \tau_- < \infty]),
\end{aligned}
$$

as $d \to \infty$. To approximate $E_0(N_0)$, we have

$$
\begin{aligned}
E_0(N_0) &= E_0^*(N_0 e^{-S_{N_0}}) \\
&\to E_0^*[\tau_- e^{-S_{\tau_-}}; \tau_- < \infty],
\end{aligned}
$$

as $d \to \infty$. Thus, we have following approximation:

Lemma 8.2: *As $d \to \infty$,*

$$
\begin{aligned}
ARL_0 &= e^d \frac{E_0^*(N_0 e^{-S_{N_0}})}{P_0^*(S_{N_0} > d) E_0^* \left(e^{-(S_{N_0}-d)} | S_{N_0} > d\right)} \\
&\approx e^d \frac{E_0^*[\tau_- e^{-S_{\tau_-}}; \tau_- < \infty]}{P_0^*(\tau_- = \infty)\nu_1}, \tag{8.4}
\end{aligned}
$$

where

$$
\begin{aligned}
\nu_1 &= \lim_{d \to \infty} E_0^* \left[e^{-(S_{N_0}-d)} | S_{N_0} > d\right] \\
&= \frac{e^{-r_1} - E_0^* \left[e^{-r_1(H_{\tau_-})}; \tau_- < \infty\right]}{P_0^*(\tau_- = \infty)}.
\end{aligned}
$$

Now we evaluate ARL_1. Similar to ARL_0, we have

$$ARL_1 = \frac{E_\mu(N_0)}{P_\mu(S_{N_0} > d)}. \tag{8.5}$$

From Robbins and Siegmund (1974), we know that

$$\left\{\sum_{n=1}^{m} \mu_{n-1}(X_n - \mu)\right\}$$

is a martingale. Thus,

$$E_\mu\left[\sum_{n=1}^{N_0} \mu_{n-1} X_n\right] = \mu E_\mu\left(\sum_{n=1}^{N_0} \mu_{n-1}\right),$$

which is equivalent to

$$
\begin{aligned}
E_\mu(N_0) &= \frac{2}{\mu^2}\left[E_\mu\left(\sum_{n=1}^{N_0} \mu_{n-1}\left(X_n - \frac{1}{2}\mu_{n-1}\right)\right) + \frac{1}{2}E_\mu\left(\sum_{n=1}^{N_0}(\mu_{n-1} - \mu)^2\right)\right] \\
&= \frac{2}{\mu^2}\left[E_\mu S_{N_0} + \frac{1}{2}E_\mu\left(\sum_{n=1}^{N_0}(\mu_{n-1} - \mu)^2\right)\right] \\
&= \frac{2}{\mu^2}[E_\mu(S_{N_0}|S_{N_0} > d)P_\mu(S_{N_0} > d) + E_\mu(S_{N_0}; S_{N_0} \le 0) \\
&\quad + \frac{1}{2}E_\mu\left(\sum_{n=1}^{N_0}(\mu_{n-1} - \mu)^2\right)].
\end{aligned}
$$

Thus, we have the following result:

Lemma 8.3: *As $d \to \infty$,*

$$
\begin{aligned}
ARL_1 &= \frac{2}{\mu^2}E_\mu(S_{N_0}|S_{N_0} > d) + \frac{2}{\mu^2}\frac{E_\mu[S_{N_0}; S_{N_0} \le 0] + \frac{1}{2}E_\mu\left(\sum_{n=1}^{N_0}(\mu_{n-1} - \mu)^2\right)}{P_\mu(S_{N_0} > d)} \\
&\approx \frac{2}{\mu^2}(d + \rho(\mu)) + \frac{2}{\mu^2}\frac{E_\mu[S_{\tau_-}; \tau_- < \infty] + \frac{1}{2}E_\mu\left(\sum_{n=1}^{\tau_-}(\mu_{n-1} - \mu)^2\right)}{P_\mu(\tau_- = \infty)},
\end{aligned}
$$

where

$$\rho(\mu) = \lim_{d \to \infty} E_\mu[S_{N_0} - d|S_{N_0} > d].$$

Thus, the adaptive CUSUM procedure is efficient at the first-order. Unfortunately, it seems difficult to obtain local second-order expansions for the related quantities. We rely on some simulation to show the results.

For Example 8.1, we let $\delta = 1.0$ and 0.5, $t = 0.0$ and 0.5, and $d = 4.8$. Table 1 gives the simulation results with 10,000 replications, where $E_0^*\left[e^{-(S_{N_0}-d)}|.\right]$ denotes $E_0^*\left[e^{-(S_{N_0}-d)}|S_{N_0} > d\right]$.

Table 8.2 gives the corresponding ARL_1 for several typical values of μ.

For Example 8.2, we take $\delta = 0.1, 0.25$, and 0.5 with $d = 4.5$. Table 8.3 gives the simulated quantities for ARL_0 with 10,000 replications.

Table 8.1: Simulated ARL_0 for Recursive Mean Estimation

d	δ	t	$P_0^*(S_{N_0} > d)$	$E_0^*[e^{-(S_{N_0}-d)}|.]$	$E_0^*(N_0 e^{-S_{N_0}})$	ARL_0
4.8	1.0	0.0	0.519	0.395	1.885	1117.5
4.8	1.0	0.5	0.518	0.447	1.890	993.7
4.8	0.5	0.0	0.367	0.442	2.132	1596.2
4.8	0.5	0.5	0.343	0.490	2.198	1589.1

Table 8.2: Simulated ARL_1 for Recursive Mean Estimation

δ	t	μ	$P_\mu(S_{N_0} > d)$	$E_\mu(N_0)$	$E_\mu[S_{N_0} - d|S_{N_0} > d]$	ARL_1
1.0	0.0	0.50	0.119	4.786	0.542	40.32
		0.75	0.277	5.322	0.669	19.19
		1.00	0.437	5.11	0.861	11.69
1.0	0.5	0.50	0.134	5.205	0.500	38.76
		0.75	0.302	5.516	0.657	18.30
		1.00	0.482	5.348	0.837	11.10
0.5	0.0	0.50	0.121	5.423	0.521	44.74
		0.75	0.265	5.583	0.666	21.07
		1.00	0.433	5.436	0.833	12.55
0.5	0.5	0.50	0.154	6.198	0.494	40.35
		0.75	0.337	6.654	0.643	19.74
		1.00	0.511	6.096	0.788	11.92

Table 8.3: Simulated ARL_0 for Robbins–Siegmund Estimator

d	δ	$P_0^*(S_{N_0} > d)$	$E_0^*[e^{-(S_{N_0}-d)}|S_{N_0} > d]$	$E_0^*(N_0 e^{-S_{N_0}})$	ARL_0
4.5	0.50	0.376	0.435	1.943	1068.9
4.5	0.25	0.282	0.448	2.025	1444.9
4.5	0.10	0.251	0.447	1.953	1574.5

Table 8.4 also gives the corresponding simulated ARL_1s for several values of μ as in Table 8.2. By comparing Table 8.2 with Table 8.4, we see that there is very little difference between the two estimators.

8.4 Biases of Estimators

In this section, we study the biases of the change-point estimator $\hat{\nu}$ and post-change mean estimator $\hat{\mu}$.

8.4.1 Bias of $\hat{\nu}$

Theoretically, we assume $\nu \to \infty$ and then d approaches ∞ conditioning on $N > \nu$. That means, we consider the quasistationary state case.

When there is no change, the zero points of T_n consist of a renewal process. Thus, as $\nu \to \infty$, $(\nu - \nu_n, T_n)$ converges to a random variable, say (L, M), where L follows the distribution

$$P_0(L = k) = P_0(\tau_- \geq k)/E_0(\tau_-), \quad \text{for} \quad k \geq 0,$$

and given $L = k$, M follows the same distribution as S_k given $S_1 > 0, ..., S_k > 0$ and $\mu_0 = \delta$. In particular, if $L = 0$, $M = 0$.

Similar to the argument given in Chapter 2 or in Srivastava and Wu (1999), we first write

$$E^{\nu}[\hat{\nu} - \nu | N > \nu] = E^{\nu}[\hat{\nu} - \nu; \hat{\nu} > \nu | N > \nu] - E^{\nu}[\nu - \hat{\nu}; \hat{\nu} < \nu | N > \nu].$$

The event $\{\hat{\nu} > \nu\}$ is asymptotically equivalent to $\tau_{-M} < \infty$ with initial state (L, M); and given $\hat{\nu} > \nu$, $\hat{\nu} - \nu$ is equivalent to τ_{-M} plus the total length of cycles of T_n coming back to zero afterward with total expected length

$$E_{\mu}[\tau_-; \tau_- < \infty]/P_{\mu}(\tau_- = \infty).$$

On the other hand, given $\hat{\nu} < \nu$, $\nu - \hat{\nu}$ is asymptotically equal to L. Thus, we have the following results.

Lemma 8.4: *As $\nu, d \to \infty$,*

$$E^{\nu}[\hat{\nu} - \nu | N > \nu] \to E_0[E_{\mu}[\tau_{-M}; \tau_{-M} < \infty | L, M]]$$

$$+ E_0[P_{\mu}(\tau_{-M} < \infty | L, M)]\frac{E_{\mu}[\tau_-; \tau_- < \infty]}{P_{\mu}(\tau_- = \infty)} - E_0[LP_{\mu}(\tau_{-M} = \infty | L, M)].$$

8.4.2   Bias of $\hat{\mu}$

To evaluate the bias of $\hat{\mu}$, we again write

$$E^{\nu}[\hat{\mu} - \mu | N > \nu] = E^{\nu}[\hat{\mu} - \mu; \hat{\nu} > \nu | N > \nu] + E^{\nu}[\hat{\mu} - \mu; \hat{\nu} < \nu | N > \nu].$$

We concentrate on the recursive mean estimation case. Given $\hat{\nu} > \nu$, $\hat{\mu}$ is equivalent to $\mu_{N_0}(t)$ conditioning on $S_{N_0} > d$ with $\mu_0 = \delta$ and $S_0 = 0$. On the other hand, given $\hat{\nu} < \nu$, $\hat{\mu}$ is equivalent to $\mu_{N_0}(t + L)$ with initial mean

$$\mu_0 = \mu'_L = \frac{\delta L + L\bar{X}'_L}{t + L}$$

and $S_0 = M$, where L, M, and μ'_L are defined from another independent copy $\{X'_1, ..., X'_n, ...\}$ of $\{X_1, ..., X_n, ...\}$.

Therefore, as $\nu \to \infty$ and d approaches infinity,

$$E^{\nu}[\hat{\mu} - \mu | N > \nu] \to E_{\mu}[\mu_{N_0} - \mu | S_{N_0} > d] E_0[P_{\mu}(\tau_{-M} < \infty | L, M)]$$

$$+ E_0[E_{\mu}[\mu_{N_0}(t + L) - \mu | S_{N_0} > d; S_0 = M; \mu_0 = \mu'_L] P_{\mu}(\tau_{-M} < \infty | L, M)].$$

It seems difficult to derive the explicit form for the bias. So we only consider the first-order by adapting the results derived in Chapter 4 for the regular CUSUM procedure.

Lemma 8.5: *As $\nu, d \to \infty$, uniformly for μ in a compact positive interval,*

$$E^{\nu}[\hat{\mu} - \mu | N > \nu] = \frac{1}{d}\{\frac{\mu^2}{2} t(\delta - \mu) + \mu + \frac{\mu^2}{2} E_0[L(\bar{X}'_L - \mu)P_{\mu}(\tau_{-M} = \infty)]$$

$$+ \frac{\mu}{2} \frac{P_{0\mu}(\tau_{-M} < \infty)}{P_{\mu}(\tau_- = \infty)} \frac{\partial}{\partial \mu} P_{\mu}(\tau_- = \infty) + E_0\left[\frac{\partial}{\partial \mu} P_{0\mu}(\tau_{-M} = \infty)\right]\}(1 + o(1)).$$

The proof will be given in the appendix.

In the special case $t = 0$, δ has no direct effect on the biases of $\hat{\mu}$.

Finally, we conduct a simulation study for the two examples. In Table 8.5, we give the simulated average delay detection time

$$ADT = E^{\nu}[N - \nu | N > \nu]$$

in the quasistationary state, the bias of $\hat{\nu}$ and $\hat{\mu}$ for $\nu = 75$ and $\mu = 0.5, 0.75, 1.0$, respectively under the designs given in Table 8.1. Due to the time consumption in this situation, all simulations are replicated 1000 times. In Table 8.6, we give the simulated average delay detection time, the bias of $\hat{\nu}$ and $\hat{\mu}$ corresponding to the designs given in Table 8.3.

Comparing Table 8.5 with Table 8.6, we can see that Example 8.2 gives a more stable bias than Example 8.1. Also, we see that the biases of $\hat{\mu}$ are quite significant for small values of μ.

Table 8.4: Simulated ARL_1 for Robbins–Siegmund Estimator

d	δ	μ	$P_\mu(S_{N_0} > d)$	$E_\mu(N_0)$	$E[S_{N_0} - d\|S_{N_0} > d]$	ARL_1
4.5	0.50	0.5	0.1154	4.6774	0.541	40.53
		0.75	0.2749	5.2532	0.661	19.11
		1.00	0.4330	5.0882	0.841	11.75
4.5	0.25	0.50	0.1088	4.6950	0.534	43.15
		0.75	0.2425	5.0640	0.675	20.88
		1.00	0.4085	5.0254	0.835	12.54
4.5	0.10	0.50	0.0991	4.5454	0.546	45.87
		0.75	0.2330	5.0000	0.668	21.47
		1.00	0.3812	5.0365	0.828	13.31

Table 8.5: Simulated Bias at $\nu = 75$ and $d = 4.8$ for Example 8.1

t	δ	μ	ADT	$E[\hat{\nu} - \nu\|N > \nu]$	$E[\hat{\mu} - \mu\|N > \nu]$
0.0	1.0	0.50	34.04	8.060	0.463
		0.75	14.54	−1.324	0.382
		1.00	8.41	−3.163	0.296
0.5	1.0	0.50	33.15	7.520	0.442
		0.75	14.18	−0.725	0.353
		1.00	7.353	−4.147	0.248
0.0	0.5	0.50	39.125	13.794	0.441
		0.75	16.49	0.245	0.343
		1.00	10.08	−1.590	0.242
0.5	0.5	0.50	38.23	9.390	0.329
		0.75	15.52	−1.407	0.275
		1.00	9.295	−3.778	0.145

Table 8.6: Simulated Bias at $\nu = 75$ and $d = 4.5$ for Example 8.2

δ	μ	ADT	$E[\hat{\nu} - \nu \vert N > \nu]$	$E[\hat{\mu} - \mu \vert N > \nu]$
0.5	0.50	33.30	11.94	0.441
	0.75	14.98	0.63	0.360
	1.00	8.77	-1.92	0.274
0.25	0.5	38.71	14.86	0.441
	0.75	16.47	1.84	0.360
	1.00	9.85	-1.65	0.266
0.1	0.50	39.61	15.96	0.450
	0.75	18.12	3.08	0.350
	1.00	9.85	-0.90	0.260

8.5   Discussions

The adaptive CUSUM procedure can not only be flexibly applied to any post-change means case but can also be conveniently used to deal with any pre-change mean model as well. Here we look at the case of an unknown constant pre-change mean case. Let μ_0 and μ be the pre-change and post-change mean respectively. The change magnitude is $\mu - \mu_0 > 0$.

We update the estimator for μ_0 after each sequential test when it goes below 0. More specifically, with little abuse of notation, we let

$$N^{(i)} = \inf\{n > 0 : S_n^{(i)} = \sum_{j=1}^{n} \delta_{j-1}^{(i-1)} \left(X_j^{(i)} - \mu_0^{(i-1)} - \frac{\delta_{j-1}^{(i-1)}}{2} \right) \leq 0; \quad \text{or} \quad > d\}$$

with $\delta_0^{(i-1)} = \delta$ and

$$\delta_{j-1}^{(i-1)} = \mu(X_1^{(i-1)} - \mu_0^{(i-1)}, ..., X_j^{(i-1)} - \mu_0^{(i-1)})$$

and if

$$S_{N^{(i)}}^{(i)} \leq 0,$$

we update $\mu_0^{(i-1)}$ to

$$\mu_0^{(i)} = \frac{(N^{(1)} + ... + N^{(i-1)})\mu_0^{(i-1)} + X_1^{(i)} + ... + X_{N^{(i)}}^{(i)}}{N^{(1)} + ... + N^{(i)}}.$$

At $N^{(1)} + ... + N^{(K)}$, where

$$K = \inf\{i \geq 1 : S_{N^{(i)}}^{(i)} > d\},$$

an alarm is made. The change-point and post-change mean are estimated as

$$\hat{\nu} = N^{(1)} + ... + N^{(K-1)},$$
$$\hat{\mu} = \mu_0^{(K-1)} + \delta_{N^{(K)}}^{(K)},$$

where $\mu_0^{(K-1)}$ is the estimator for the pre-change mean.

8.6   Appendix

8.6.1   Limiting Distribution of W_n

Denote $k = [tn]$, for $0 \le 0 \le 1$. For Z_i i.i.d. following N(0, 1), it is known that

$$\frac{1}{\sqrt{n}}(Z_1 + \ldots + Z_{[nt]}) \Longrightarrow B(t),$$

strongly as $n \to \infty$ for any finite time interval. Thus, by writing

$$U_n(t) = (W_1 + \ldots + W_{[nt]})/\sqrt{n},$$

we see that as $n \to \infty$, $U_n(t)$, which satisfies

$$\frac{W_{[nt]}}{\sqrt{n}} = \frac{1}{t}(U_n(t) + B_n(t))^+,$$

strongly converges to $U(t)$, which satisfies

$$\frac{dU(t)}{dt} = \frac{1}{t}(U(t) + B(t))^+,$$

for $0 \le t \le 1$. Thus, W_n strongly converges to $dU(1)/dt$.

8.6.2   Proof of Lemma 8.5

The proof generalizes the results in Wu (2004b).

   Note that conditional on $S_{N_0} > d$, uniformly for μ in a compact positive interval, as $d \to \infty$,

$$N_0 = \frac{d}{\mu^2/2}(1 + o_p(1))$$

and

$$\bar{X}_{N_0} = \mu(1 + o_p(1)).$$

We write

$$\mu_{N_0} = \frac{t}{t + N_0}\delta + \frac{N_0}{t + N_0}\bar{X}_{N_0}$$

$$= \left(\frac{t(\delta - \mu)}{N_0} + \bar{X}_{N_0}\right)(1 + o_p(1)).$$

Thus, similar to the proof of Lemma 4.1 of Chapter 4, we have

$$E_\mu[\mu_{N_0} - \mu; S_{N_0} > d]$$

$$= \frac{t(\delta - \mu)}{d} \frac{\mu^2}{2} P_\mu(S_{N_0} > d) + \frac{1}{d} \frac{\partial}{\partial \mu} \left(\frac{\mu^2}{2} P_\mu(S_{N_0} > d) \right) (1 + o(1))$$

$$= \frac{1}{d} \left(\frac{\mu^2}{2} t(\delta - \mu) P_\mu(\tau_- = \infty) + \frac{\partial}{\partial \mu} \left(\frac{\mu^2}{2} P_\mu(S_{N_0} > d) \right) \right) (1 + o(1)).$$

Thus,

$$E_\mu[\mu_{N_0} - \mu | S_{N_0} > d] = \frac{1}{d} \left(\frac{\mu^2}{2} t(\delta - \mu) + \mu + \frac{\mu^2}{2} \frac{\frac{\partial}{\partial \mu} P_\mu(S_{N_0} > d)}{P_\mu(S_{N_0} > d)} \right) (1 + o(1)).$$

Similarly,

$$E_0[E_\mu(\mu_{N_0}(t + L) - \mu | S_{N_M} > d)]$$

$$= E_0 \left[E_\mu \left[\frac{t\delta + L\bar{X}'_L + N_M \bar{X}_{N_M}}{N_M + t + L} - \mu | S_{N_M} > d \right] \right]$$

$$= \frac{1}{d} \{ \frac{t\mu^2}{2}(\delta - \mu) P(\tau_{-M} = \infty) + \frac{\mu^2}{2} E[L(\bar{X}'_L - \mu); \tau_{-M} = \infty]$$

$$+ \frac{\partial}{\partial \mu} \left(\frac{\mu^2}{2} P(\tau_{-M} = \infty) \right) \} (1 + o(1)).$$

Combining the above approximations, we get the expected results.

8.7 Case Study

8.7.1 Global Warming Detection

The data set in Table 8.7 gives the global average temperature from 1855–1997 as an illustration to detect the locations and rates of increment. The data are read in rows.

The same data have been treated in Wu, Woodroofe, and Mentz (2000) by fitting an isotonic regression model. A piecewise linear model is used in Karl, Knight, and Baker (2000) for a smaller data set (1880–1997) by fitting an AR(1) time series error model with wavelet analysis. It reveals that there are increment periods.

The scatterplot of the global average temperatures $\{y_i\}$ for $i = 1, ..., 142$ (years 1855–1997) shows that there are two increment periods and for each increment period, the change is roughly linear.

To detect the first change, we use the initial value $\mu_0 = -0.32$, which is the mean of the first 30 observations with s.d. $= 0.11$. Then the data are standardized as

$$x_i = (y_i + 0.32)/0.11,$$

Table 8.7: Global Average Temperature (1855–1997)

−0.36	−0.48	−0.43	−0.25	−0.39	−0.43	−0.51	−0.29
−0.47	−0.27	−0.21	−0.31	−0.24	−0.30	−0.34	−0.36
−0.23	−0.29	−0.41	−0.42	−0.40	−0.13	0.00	−0.30
−0.28	−0.25	−0.23	−0.31	−0.36	−0.34	−0.27	−0.37
−0.31	−0.17	−0.39	−0.32	−0.42	−0.46	−0.38	−0.36
−0.16	−0.15	−0.33	−0.22	−0.13	−0.22	−0.37	−0.44
−0.49	−0.37	−0.30	−0.50	−0.52	−0.49	−0.46	−0.48
−0.41	−0.42	−0.24	−0.13	−0.36	−0.51	−0.39	−0.30
−0.23	−0.19	−0.30	−0.27	−0.33	−0.22	−0.08	−0.19
−0.22	−0.37	−0.13	−0.05	−0.10	−0.23	−0.11	−0.15
−0.10	0.00	0.10	0.02	−0.04	0.06	0.06	0.06
0.22	0.06	−0.08	−0.08	−0.08	−0.09	−0.19	−0.05
0.02	0.10	−0.15	−0.16	−0.26	0.05	0.12	0.04
0.00	0.03	0.04	0.07	−0.22	−0.16	−0.06	−0.06
−0.09	0.03	−0.03	−0.19	−0.06	0.08	−0.18	−0.12
−0.22	0.06	−0.03	0.06	0.10	0.14	0.05	0.24
0.02	0.00	0.09	0.23	0.25	0.18	0.35	0.29
0.15	0.19	0.26	0.39	0.22	0.43		

which are assumed to be i.i.d. $N(0,1)$ r.v.s.

For two initial values $\beta_0 = 0.25$ and 0.5 with $d = 30$, the adaptive CUSUM procedures with recursive least-square estimation for β give the same detection time $N = 89$ (which is the year 1944), and $\hat{\nu} = 74$ (which is the year 1929) with 15 delayed observations. The change slope is thus estimated as

$$\hat{\beta} = \sum_{j=75}^{89} \left(\frac{X_j + 0.32}{0.11} \right) (j - 74) / \sum_{j=75}^{89} (j - 74)^2 = 0.301.$$

So without the bias correction, the fitted model becomes

$$x_i = 0.301(i - 74)^+ + \epsilon_i,$$

or

$$y_i = -0.32 + 0.0331(i - 74)^+ + 0.11\epsilon_i,$$

for $i = 1, ..., 89$.

To detect the second increment, we start from $i = 90$ (which is guessed without formal detection) with initial mean $\mu_0 = -0.054$, which is the average of $y_{91}, ..., y_{120}$ with s.d. $= 0.10$. The data are then transformed into

$$x_i = (y_i + 0.054)/0.10.$$

With $d = 30$, and two initial values $\beta_0 = 0.25$ and 0.5, the adaptive CUSUM procedure gives the same detection time $N = 139$ (which is the year 1994) and the change point $\hat{\nu} = 123$ (which is the year 1978). The change slope is then estimated as $\hat{\beta} = 0.227$.

The fitted model without bias correction is

$$y_i = -0.054 + 0.0227(i - 123)^+ + 0.10\epsilon_i,$$

for $i = 90, ..., 142$.

The residual plotting shows that the fitted model is quite good.

9
Dependent Observation Case

9.1 Introduction

So far, we have concentrated our discussion on the independent observation case. When the observations become dependent, it will affect the performance of the CUSUM procedure.

Lai (1995) discussed the detection of change-point in general dynamic systems when the observations follow a time series model and proposed a window-limited version of the generalized likelihood ratio detection procedure. Further developments are referred to Lai (1998, 2002) and Lai and Shan (1999).

The effect of correlation on the performance of operating characteristics for the CUSUM procedure has been studied by a number of authors, such as Johnson and Bagshaw (1974), Bagshaw and Johnson (1975), Yashchin (1993), Schmid (1997), Fuh (2003).

There are basically two methods to deal with the correlated data.

When the dynamic structure of the model is complicated or not clear, or if the primary goal is to detect changes of distributional parameters such as the mean or variance, we would like to use the classical CUSUM procedure and make corresponding adjustments on the design of the threshold. The method is obviously not optimal under specific model assumptions, but it does not lose the direct meaning of explanation. The disadvantage is that the theoretical evaluation becomes much more complicated; see Fuh (2003) in the Markovian structure case.

When the dynamic structure of the model is clear and not very complicated, or if the primary goal is to detect changes of structural parameters, such as the correlation coefficients, then a model-based CUSUM procedure is preferred. This method has the advantage that the formed CUSUM procedure is typically based on the innovation (residual) process, which becomes independent in the normal case. However, the post-change structure of the innovation process becomes complicated and needs more careful treatment based on the model assumption.

Our primary goal here is to study the effect of dependence under the two methods in terms of change-point estimation and post-change parameter estimation. We shall concentrate our emphasis on AR(1) model due to the following several reasons.

First, the AR(1) model captures the first-order dependence, which is typically used for modelling the dependence at the first step. Although higher-order dependence is always possible for real data set, the method can be extended to general models as well.

Second, the AR(1) model is a typical example to study the change in either distributional parameters or structural parameters since it has a simple form of dynamic structure.

Third, exact analysis can be done under the AR(1) model as we shall show in the following sections, which can demonstrate both qualitatively and quantitatively the effect of the correlation.

In the following two sections, we shall study in detail the effect of the correlation on the bias of change-point estimator and post-change mean under the AR(1) model. Exact theoretical results will be given for the model-based CUSUM procedure. Then the comparison will be made with the classical CUSUM procedure.

9.2 Model-based CUSUM Procedure

9.2.1 AR(1) Change-Point Model

We assume the original observations $\{Y_i\}$ follow the model

$$Y_i - \mu_i = \phi(Y_{i-1} - \mu_{i-1}) + \epsilon_i,$$

for $i = 1, 2, ...$, where Y_0 follows $N(\mu_0, 1/(1 - \phi^2))$ in the stationary case and $|\phi| < 1$ is the autocorrelation. Assume for $i \leq \nu$, $\mu_i = \mu_0 < 0$ and for $i > \nu$, $\mu_i = \mu > 0$, and ϵ_i are i.i.d. standard normal random variables.

Assume $\mu_1 = -\mu_0 > 0$ is the reference value for μ.

If we ignore the effect of the correlation, we can still use the classical CUSUM procedure by treating μ_0 as θ_0 and μ as θ. The study of the biases for the change-point estimator and post-change mean estimator becomes complicated even under this simple model. Only the theoretical results can be obtained as in Fuh (2003). Simulation results are reported in the next section.

In the following, we consider the model-based CUSUM procedure based on the likelihood ratio approach.

We assume ϕ is known and $|\phi| < 1$.

Define

$$X_i = Y_i - \phi Y_{i-1},$$

and $\theta = (1 - \phi)\mu$ and $\theta_i = (1 - \phi)\mu_i$ for i=0,1, respectively.

Then, we see that the model-based CUSUM procedure is formed by X_i's and defined as

$$T_n = \max(0, T_{n-1} + X_n),$$

for $T_0 = 0$. A signal of change is made at

$$N = \inf\{n > 0 : T_n > d\}.$$

However, we note that for $i \leq \nu$, the mean of X_i is θ_0 and for $i > \nu + 1$, the mean of X_i is θ_1; while at $i = \nu + 1$, the mean of $X_{\nu+1}$ is

$$\theta_1 + \beta(\theta_1 - \theta_0),$$

where $\beta = \phi/(1 - \phi)$.

Thus, we see that the model-based CUSUM procedure does not follow the conventional model assumptions.

9.2.2 Bias of Change-Point Estimator

We shall use the same notations as in Chapter 2. We assume ν and d approach ∞ and consider the bias of $\hat{\nu}$ conditional on $N > \nu$. When a change occurs, T_n is asymptotically equivalent to M. $X_{\nu+1}$ has mean $\theta_1 + \beta(\theta - \theta_0)$. Thus, the event that the CUSUM process directly crosses the boundary d without going back to zero is equivalent to the event that the random walk S_n starts at $M + \beta(\theta - \theta_0)$ and then directly crosses the boundary d without going below zero. Therefore, we have the following result:

Theorem 9.1 : *As $\nu, d \to \infty$,*

$$E^\nu[\hat{\nu} - \nu | N > \nu] \to E_{\theta_0\theta}[\tau_{-(M+\beta(\theta_1-\theta_0))}; \tau_{-(M+\beta(\theta-\theta))} < \infty]$$

$$+ P_{\theta_0\theta}(\tau_{-(M+\beta(\theta-\theta_0))} < \infty)E_\theta(\gamma_m) - E_{\theta_0\theta}[\sigma_M; \tau_{-(M+\beta(\theta-\theta_0))} = \infty],$$

$$E^\nu[|\hat{\nu} - \nu| | N > \nu] \to E_{\theta_0\theta}[\tau_{-(M+\beta(\theta-\theta_0))}; \tau_{-(M+\beta(\theta-\theta_0))} < \infty]$$

$$+ P_{\theta_0\theta}(\tau_{-(M+\beta(\theta-\theta_0))} < \infty)E_\theta\gamma_m) + E_{\theta_0\theta}[\sigma_M; \tau_{-(M+\beta(\theta-\theta_0))} = \infty].$$

Now, we further assume $\theta_0, \theta \to 0$ at the same order and derive a second-order approximation for the asymptotic bias and absolute bias.

The derivation is similar to Chapter 2, but needs some careful technical treatments. We give only the main steps. Since we are only dealing with the normal distribution case, the notations can be simplified due to the symmetric property. For example, $\tilde{\theta} = -\theta$ and $\theta_1 = -\theta_0$, and $\Delta_0 = -2\theta_0$ and $\Delta = 2\theta$. Finally, we note that $\mu = \theta$ and $\mu_0 = \theta_0$.

First, we note that

$$E_\theta(\gamma_m) = \frac{1}{2\theta^2} - \frac{1}{4} + o(1).$$

To evaluate $P_{\theta_0\theta}(\tau_{-(M+\beta(\theta-\theta_0))} < \infty)$, we follow a similar technique used in Chapter 2.

By conditioning on whether $M = 0$ or $M > 0$, we have

$$P_{\theta_0\theta}(\tau_{-(M+\beta(\theta-\theta_0))} < \infty)$$

$$= P_\theta(\tau_{-\beta(\theta-\theta_0)} < \infty)P_{\theta_0}(\tau_+ = \infty) + P_{\theta_0\theta}(\tau_{-(M+\beta(\theta-\theta_0))} < \infty; M > 0). \quad (9.1)$$

By using Wald's likelihood ratio identity, we have

$$P_\theta(\tau_{-\beta(\theta-\theta_0)} < \infty) = E_{-\theta}e^{2\theta S_{\tau_{-\beta(\theta-\theta_0)}}}$$
$$= 1 + 2\theta E_0 S_{\tau_-} + o(\theta).$$

Thus, from Lemma 1.2, we have

$$P_\theta(\tau_{-\beta(\theta-\theta_0)} < \infty)P_{\theta_0}(\tau_+ = \infty) = -2\theta_0 E_0 S_{\tau_+} e^{\theta_0 \rho_+}(1 + 2\theta E_0 S_{\tau_+}) + o(\theta^2)$$
$$= -2\theta_0 E_0 S_{\tau_+} e^{\theta_0 S_{\tau_+}} + 2\theta_0\theta + o(\theta^2).$$

For the second term of (9.1), by using the Wald's likelihood ratio identity by changing parameter θ to $-\theta$ and θ_0 to θ_1, we have

$$P_\theta(\tau_x < \infty) = E_{-\theta}e^{2\theta S_{\tau_{-x}}}$$
$$= e^{-2\theta x}E_{-\theta}e^{2\theta R_{-x}}$$

and

$$P_{\theta_0}(M > x) = P_{\theta_0}(\tau_x < \infty)$$
$$= e^{2\theta_0 x}E_{\theta_1}e^{2\theta_0 R_x}.$$

Now, we write

$$P_{\theta_0\theta}(\tau_{-(M+\beta(\theta-\theta_0))} < \infty, M > 0)$$

$$= -\int_0^\infty P_\theta(\tau_{-(x+\beta(\theta-\theta_0))} < \infty)dP_{\theta_0}(M > x)$$

$$= -\int_0^\infty E_{-\theta}e^{2\theta S_{\tau_{-(x+\beta(\theta-\theta_0))}}} dE_{\theta_1}e^{2\theta_0 S_{\tau_x}}$$

$$= -\int_0^\infty e^{-2\theta(x+\beta(\theta-\theta_0)-\rho_-)}de^{2\theta_0(x+\rho_+)}$$

$$-\int_0^\infty e^{-2\theta(x+\beta(\theta-\theta_0)-\rho_-)}d\left(e^{2\theta_0(x+\rho_+)}(E_{\theta_1}e^{2\theta_0(R_x-\rho_+)})-1\right)$$

$$-\int_0^\infty e^{-2\theta(x+\beta(\theta-\theta_0)-\rho_-)}\left(E_{\theta_0}e^{-2\theta(R_{-(x+\beta(\theta-\theta_0))}-\rho_-)}-1\right)de^{2\theta_0(x+\rho_+)}$$

$$-\int_0^\infty e^{-2\theta(x+\beta(\theta-\theta_0)-\rho_-)}\left(E_{\theta_0}e^{2\theta(R_{-(x+\beta(\theta-\theta_0))}-\rho_-)}-1\right)$$

$$\times d\left(e^{2\theta_0(x+\rho_+)}(E_{\theta_1}e^{2\theta_0(R_x-\rho_+)})-1\right). \tag{9.2}$$

The first term of (9.2) is

$$\frac{-\theta_0}{\theta-\theta_0}e^{2\theta(-\beta(\theta-\theta_0)+\rho_-)+2\theta_0\rho_+}.$$

The third term of (9.2) is approximately equal to

$$-4\theta_0\theta\int_0^\infty (E_0R_{-x}-\rho_-)dx + o(\theta^2).$$

The fourth term of (9.2) is

$$-4\theta_0\theta\int_0^\infty E_0(R_{-x}-\rho_-)dE_0(R_x-\rho_+) + o(\theta^2).$$

The second term of (9.2), by integrating by parts, can be approximated as

$$e^{2\theta(-\beta(\theta-\theta_0)+\rho_-)}\left(P_{\theta_0}(\tau_+<\infty)-e^{2\theta_0\rho_+}\right)-4\theta_0\theta\int_0^\infty(E_0R_x-\rho_+)dx+o(\theta^2).$$

Combining the above approximations, we have

Lemma 9.1: *As $\theta_0, \theta \to 0$ at the same order,*

$$P_{\theta_0\theta}(\tau_{-(M+\beta(\theta-\theta_0))}<\infty)=\frac{-\theta_0}{\theta-\theta_0}+\theta\theta_0+2\beta\theta\theta_0+o(\theta^2).$$

Thus,

$$E_\theta(\gamma_m)P_{\theta_0\theta}(\tau_{-(M+\beta(\theta-\theta_0))}<\infty)=\frac{-\theta_0}{2\theta^2(\theta-\theta_0)}+\frac{\theta_0}{2\theta}+\frac{\theta_0}{4(\theta-\theta_0)}+\beta\frac{\theta_0}{\theta}+o(1),$$

with an extra term $\beta\frac{\theta_0}{\theta}$ compared with the one of Chapter 2 in the normal case. Next, depending on $M = 0$ or $M > 0$, we write

$$E_{\theta_0\theta}[\tau_{-(M+\beta(\theta-\theta_0))}; \tau_{-(M+\beta(\theta-\theta_0))}<\infty]=E_\theta[\tau_{-\beta(\theta-\theta_0)}; \tau_{-\beta(\theta-\theta_0)}<\infty]$$

$$-\int_0^\infty E_\theta[\tau_{-(x+\beta(\theta-\theta_0))}; \tau_{-(x+\beta(\theta-\theta_0))}<\infty]dP_{\theta_0}(\tau_x<\infty). \tag{9.3}$$

As $\theta_0, \theta \to 0$, the first term of (9.3) can be shown to be

$$E_\theta[\tau_-; \tau_- < \infty]P_{\theta_0}(\tau_+ = \infty) = -\frac{\theta_0}{\theta} + o(1),$$

as in Chapter 2. For the second term of (9.3), we write it as

$$-\int_0^\infty E_{-\theta}\left[\tau_{-(x+\beta(\theta-\theta_0))}e^{-2\theta(x+\beta(\theta-\theta_0))+R_{-(x+\beta(\theta-\theta_0))}}\right]dE_{-\theta_0}e^{2\theta_0(x+R_x)}$$

$$= -\int_0^\infty E_{-\theta}(\tau_{-(x+\beta(\theta-\theta_0))})e^{-2\theta(x+\beta(\theta-\theta_0))+\rho_-}dE_{-\theta_0}e^{2\theta_0(x+R_x)} + o(1)$$

$$= \frac{1}{\theta}\int_0^\infty E_{-\theta}S_{\tau_{-(x+\beta(\theta-\theta_0))}}dE_{-\theta_0}e^{2\theta_0(x+R_x)} + o(1)$$

$$= \frac{1}{\theta}[\int_0^\infty (-(x+\beta(\theta-\theta_0))+E_0R_{-x})e^{-2\theta(x+\beta(\theta-\theta_0))+\rho_-}$$

$$- 2\theta_0\int -0^\infty(-(x+\beta(\theta-\theta_0))+E_0R_{-x})d(E_0R_x - \rho_+)] + o(1)$$

$$= -\frac{2\theta_0}{\theta}\{e^{2\theta\rho_-+2\theta_0\rho_+-2\beta\theta(\theta-\theta_0)}\left(\frac{\beta(\theta-\theta_0)}{2(\theta-\theta_0)} + \frac{1}{4(\theta-\theta_0)^2}\right)$$

$$+ \rho_-(E_0S_{\tau_+} - \rho_+) - \int_0^\infty (E_0R_{-x} - \rho_-)d(E_0R_x - \rho_+)\} + o(1).$$

Comparing the above form with the corresponding form in Chapter 2, we see that two extra terms coming out up to the order $o(1)$. One term is equal to

$$-\frac{2\theta_0}{\theta}\frac{\beta(\theta-\theta_0)}{2(\theta-\theta_0)} = -\frac{\theta_0}{\theta}\beta,$$

and the other term is

$$-\frac{2\theta_0}{\theta}-2\beta\theta(\theta-\theta_0)\frac{1}{4(\theta-\theta_0)^2} = \beta\frac{\theta_0}{\theta-\theta_0}.$$

Finally, we look at the term

$$E_{\theta_0\theta}[\sigma_M; \tau_{-(M+\beta(\theta-\theta_0))} = \infty]$$

$$= E_{\theta_0}\sigma_M - E_{\theta_0}[\sigma_M P_\theta(\tau_{-(M+\beta(\theta-\theta_0))} < \infty)]$$

$$= E_{\theta_0}\sigma_M - E_{\theta_0}\left[\sigma_M E_{-\theta}\left[\exp(2\theta(-(M+\beta(\theta-\theta_0))+R_{-(M+\beta(\theta-\theta_0))}))\right]\right]$$

$$= E_{\theta_0}\sigma_M - E_{\theta_0}\left[\sigma_M e^{-2\theta(M+\beta(\theta-\theta_0))+\rho_-}\right](1+o(\theta^2)).$$

We see that the only difference comes from the term $\exp(-2\theta\beta(\theta-\theta_0))$. Since

$$E_{\theta_0}\sigma_M = \frac{1}{2\theta_0^2} - \frac{1}{4} + o(1)$$

and

$$E_{\theta_0}\left[\sigma_M e^{-2\theta(M+\beta(\theta-\theta_0))+\rho_-}\right] = -\frac{1}{2(\theta-\theta_0)^2} - \frac{1}{4} - \beta\frac{\theta}{\theta-\theta_0} + o(1),$$

we have,

$$E_{\theta_0\theta}[\sigma_M;\tau_{-(M+\beta(\theta-\theta_0))} = \infty] = \frac{1}{2\theta_0^2} - \frac{1}{2(\theta-\theta_0)^2} + \beta\frac{\theta}{\theta-\theta_0} + o(1),$$

with an extra term $\beta\frac{\theta}{\theta-\theta_0}$.

Finally, by combining the above approximations, we have the following result:

Corollary 9.1: *As* $\theta, \theta_0 \to 0$ *at the same order,*

$$E^\nu[\hat\nu - \nu|N > \nu] \approx \frac{1}{2\theta^2} - \frac{1}{2\theta_0^2} + \frac{\theta_0}{4(\theta-\theta_0)} - \beta + o(1),$$

$$E^\nu[|\hat\nu - \nu||N > \nu] \approx \frac{1}{2\theta^2} + \frac{1}{2\theta_0^2} - \frac{1}{(\theta-\theta_0)^2} + \frac{\theta_0}{4(\theta-\theta_0)} + \beta\frac{\theta+\theta_0}{\theta-\theta_0} + o(1).$$

When $\theta = -\theta_0$, *we have*

$$E^\nu[\hat\nu - \nu|N > \nu] \approx -\frac{1}{8} - \beta + o(1),$$

$$E^\nu[|\hat\nu - \nu||N > \nu] \approx \frac{3}{4\theta_0^2} - \frac{1}{8} + o(1).$$

We see that when $\theta = -\theta_0$, the correlation has an effect on the bias at the second-order and when the correlation is small, the effect is almost negligible. However, when $\theta \neq -\theta_0$ or $\mu \neq -\mu_0$, the effect of the correlation can be significant as $\theta = (1-\phi)\mu$ and thus there is a factor $1/(1-\phi)^2$ at the first-order.

Remark: The analysis for the case $\phi < 0$ is technically different, although the results are the same. The main difference comes from the fact that when $\beta < 0$, at $M = 0$, the notation $\tau_{-\beta(\theta-\theta_0)}$ will have a different meaning since $-\beta(\theta - \theta_0) > 0$. We shall not pursue this in detail here.

9.2.3   Bias of $\hat\theta$

Again, we constrain our discussion to the case $\phi \geq 0$ for notational and technical convenience.

By separating the event $\{\hat\nu \leq \nu\}$ with $\{\hat\nu > \nu\}$, we first write

$$E^\nu[\hat\theta - \theta|N > \nu] = E^\nu[\hat\theta - \theta; \hat\nu \leq \nu|N > \nu] + E^\nu[\hat\theta - \theta; \hat\nu > \nu|N > \nu].$$

Note that

$$P^\nu(\hat\nu > \nu) \to P_{\theta_0\theta}(\tau_{-(M+\beta(\theta-\theta_0))} < \infty).$$

Given $\{\hat{\nu} > \nu\}$, $\hat{\theta}$ is equivalent to S_{N_0}/N_0 in distribution as in Chapter 4. Thus, as $\nu \to \infty$ and d approaches infinity,

$$E^{\nu}[\hat{\theta} - \theta | \hat{\nu} > \nu] \quad \to \quad E_{\theta}\left[\frac{S_{N_0}}{N_0} | S_{N_0} > d\right]$$

$$= \frac{a_0(\theta)}{d}(1 + o(1)),$$

where

$$a_0(\theta) = 1 + \theta\frac{\partial}{\partial\theta}\ln P_{\theta}(\tau_- = \infty).$$

On the other hand, note that the event $\{\hat{\theta} \leq \nu\}$ is equivalent to $\{S_{N_{M+\beta(\theta-\theta_0)}} > d\}$, and $\hat{\theta}$ is equivalent to

$$\frac{S_{N_{M+\beta(\theta-\theta_0)}}}{N_{M+\beta(\theta-\theta_0)} + \sigma_M}.$$

Thus, as $\nu \to \infty$, we have

$$E^{\nu}[\hat{\theta} - \theta; \hat{\nu} \leq \nu | N > \nu]$$

$$\to \quad E_{\theta_0\theta}\left[\frac{S_{N_{M+\beta(\theta-\theta_0)}}}{N_{M+\beta(\theta-\theta_0)} + \sigma_M} - \theta; S_{N_{M+\beta(\theta-\theta_0)}} > d\right]$$

$$= \quad E_{\theta_0\theta}\left[\frac{S_{N_{M+\beta(\theta-\theta_0)}} - (M + \beta(\theta - \theta_0))}{N_{M+\beta(\theta-\theta_0)} + \sigma_M} - \theta; S_{N_{M+\beta(\theta-\theta_0)}} > d\right]$$

$$+ E_{\theta_0\theta}\left[\frac{M + \beta(\theta - \theta_0)}{N_{M+\beta(\theta-\theta_0)} + \sigma_M}; S_{N_{M+\beta(\theta-\theta_0)}} > d\right].$$

For the first term on the right-hand side, we can follow the same line as in Chapter 4 by noting that

$$\hat{\theta}_{M+\beta(\theta-\theta_0)} = \frac{S_{N_{M+\beta(\theta-\theta_0)}} - (M + \beta(\theta - \theta_0))}{N_{M+\beta(\theta-\theta_0)}},$$

and it can be shown to be

$$\frac{1}{d}E_{\theta_0}\left[\frac{\partial}{\partial\theta}(\theta P_{\theta}(\tau_{-(M+\beta(\theta-\theta_0))} = \infty))\right] - \frac{\theta^2}{d}E_{\theta_0\theta}[\sigma_M; \tau_{-(M+\beta(\theta-\theta_0))} = \infty](1 + o(1)).$$

Similarly, the second term can be shown as

$$\frac{\theta}{d}E_{\theta_0\theta}[M + \beta(\theta - \theta_0); \tau_{-(M+\beta(\theta-\theta_0))} = \infty].$$

Combining all the above results, we get the following first-order asymptotic bias for $\hat{\theta}$.

Theorem 9.2: *As $\nu \to \infty$ and d approaches infinity,*

$$E^\nu[\hat{\theta} - \theta | N > \nu] = \frac{1}{d}\{1 + \theta\frac{\partial}{\partial\theta}\ln P_\theta(\tau_- = \infty)P_{\theta_0\theta}(\tau_{-(M+\beta(\theta-\theta_0))}) < \infty)$$

$$+ \theta\frac{\partial}{\partial\theta}P_{\theta_0\theta}(\tau_{-(M+\beta(\theta-\theta_0))}) = \infty) + \theta E_{\theta_0\theta}[(M - \theta\sigma_M); \tau_{-(M+\beta(\theta-\theta_0))} = \infty]$$

$$+ \beta\theta(\theta - \theta_0)P_{\theta_0\theta}(\tau_{-(M+\beta(\theta-\theta_0))}) = \infty)\}(1 + o(1)).$$

A detailed local second-order analysis as θ_0, θ approach zero shows that the effect of β is only at the second order in form. Thus we have the following same formula for the local approximation:

Corollary 9.2: *If $\theta_0, \theta \to 0$ at the same order,*

$$E^\nu[\hat{\theta} - \theta | N > \nu] \approx \frac{1}{d}\left[2 - \frac{\theta^3}{2\theta_0^2(\theta - \theta_0)}\right](1 + o(1)),$$

which is equivalent to

$$E^\nu[\hat{\mu} - \mu | N > \nu] \approx \frac{1}{(1-\phi)d}\left[2 - \frac{\mu^3}{2\mu_0^2(\mu - \mu_0)}\right](1 + o(1)).$$

Thus, the effect on the bias for the estimation for μ will be proportional to $1/(1-\phi)$. Positive ϕ's inflate the bias.

9.3 Numerical Results

In this section, we conduct some numerical studies to serve two purposes. First, we demonstrate the theoretical results in the model-based CUSUM procedure under the AR(1) model. Second, we check the effect of correlation on the change-point and post-change mean estimators under the classical CUSUM procedure.

Table 9.1 gives some simulation results for $\nu = 50, 100$ and $\mu_0 = -\mu = -0.25, 0.5$ with $d = 10$. For $\phi = 0.05, 0.1$, and 0.2, we give the simulated bias for the change-point estimator $\hat{\nu}$ and post-change parameter estimator $\hat{\theta}$ based on 1000 replications, as in Table 2.1 of Chapter 2. For the AR(1) process, we assume Y_0 follows the stationary distribution $N(0, 1/(1 - \phi^2))$.

From the table, it seems that due to the large variance, the simulated bias for $\hat{\nu}$ is not very reliable. However, the simulated bias for the post-change parameter estimator $\hat{\theta}$ is quite satisfactory (although the simulated errors are not reported due to the similarity to Chapter 4).

Table 9.2 gives some simulated biases for the change-point estimator and post-change mean estimator by using the classical CUSUM procedure. Again, we take $\nu = 50, 100$ with $d = 10$. For several typical values of ϕ, we use 1000 replications with Y_0 following the stationary distribution $N(0, 1/(1 - \phi^2))$.

It seems that under the classical CUSUM procedure, the bias for the change-point estimator gets larger as the correlation gets more positive, so does the

Table 9.1: Simulated Bias under Model-based CUSUM Procedure

μ_0	ν	ϕ	$\theta_0 = -\theta$	$P(N > \nu)$	$E[\hat{\nu} - \nu \mid N > \nu]$	$E[\hat{\theta} - \theta \mid N > \nu]$
−0.25	50	0.05	−0.238	0.989	0.627	0.181
		0.10	−0.225	0.990	1.541	0.181
		0.20	−0.200	0.973	1.853	0.174
	100	0.05	−0.238	0.959	0.892	0.165
		0.10	−0.225	0.949	0.456	0.174
		0.20	−0.200	0.924	1.795	0.190
−0.50	50	0.05	−0.475	1.000	−0.450	0.136
		0.10	−0.450	1.000	−0.352	0.135
		0.20	−0.400	0.997	−0.297	0.161
	100	0.05	−0.475	0.999	−0.024	0.146
		0.10	−0.450	0.999	0.093	0.143
		0.20	−0.400	0.997	−0.100	0.151

bias for the post-change mean estimator. However, we have to note that the design of the threshold d depends on the correlation ϕ. So although it has very little effect on the bias of change-point estimator, it may affect the sample size $N - \hat{\nu}$ and thus affect the bias of the post-change estimator. A systematic simulation study or a theoretical verification will be necessary to conduct a full comparison between the classical CUSUM procedure and the model-based CUSUM procedure in terms of the change-point and post-change parameter estimation.

Table 9.2: Bias under Classical CUSUM Procedure

ν	$\mu_0 = -\mu$	ϕ	$P(N > \nu)$	$E[\hat{\nu} - \nu \mid N > \nu]$	$E[\hat{\mu} - \mu \mid N > \nu]$
50	−0.25	−0.20	0.998	0.336	0.122
		−0.10	0.992	1.124	0.142
		−0.05	0.996	0.801	0.160
	−0.25	0.05	0.981	1.766	0.188
		0.10	0.965	2.568	0.212
		0.20	0.921	2.762	0.242
	−0.5	−0.20	1.000	−0.173	0.103
		−0.10	1.000	0.040	0.120
		−0.05	1.000	0.074	0.149
		0.05	0.998	−0.276	0.135
		0.10	1.000	−0.344	0.167
		0.20	0.996	0.088	0.183
100	−0.25	−0.20	0.995	0.336	0.127
		−0.10	0.978	0.660	0.140
		−0.05	0.976	−0.772	0.149
		0.05	0.940	1.000	0.188
		0.10	0.909	1.190	0.212
		0.20	0.840	1.538	0.250
	−0.50	−0.20	1.000	0.016	0.114
		−0.10	0.999	−0.150	0.124
		−0.05	0.999	−0.005	0.146
		0.05	1.000	−0.262	0.148
		0.10	0.998	0.064	0.155
		0.20	0.992	0.218	0.207

10
Other Methods and Remarks

10.1   Shiryayev−Roberts Procedure

For detecting change-point, several other procedures are available such as the Shewhart chart, EWMA chart, and Shiryayev−Roberts procedure. The CUSUM procedure has been the focus in our discussion due to its interpretation from generalized likelihood ratio test. In this section, we consider the inference problem for the change-point and post-change mean from a Bayesian point of view by using the Shiryayev−Roberts procedure as the detecting procedure.

The Shiryayev−Roberts procedure has been further discussed by Pollak (1985, 1987) for its optimality and design in the discrete-time case. Here we use the simple formula given in Wu (1994) for design purposes.

We assume that the change-point follows a geometric prior distribution with

$$P(\nu = k) = (1 - p)^{k-1}p, \quad \text{for} \quad k = 1, 2, ...$$

Given data $x_1, ..., x_n$, the posterior distribution of ν can be written as

$$
\begin{aligned}
\pi_{k,n} &= P(\nu = k | x_1, ..., x_n) \\
&= \frac{f(x_1, ..., x_n | \nu = k)P(\nu = k)}{\sum_{j=1}^{\infty} f(x_1, ..., x_n | \nu = j)P(\nu = j)}
\end{aligned}
$$

$$= \frac{\Pi_{i=1}^{k} f_{\theta_0}(x_i) \Pi_{i=k+1}^{n} f_{\theta_1}(x_i)(1-p)^{k-1}p}{\sum_{j=1}^{n} \Pi_{i=1}^{j} f_{\theta_0}(x_i) \Pi_{i=j+1}^{n} f_{\theta_1}(x_i)(1-p)^{j-1}p + (1-p)^n \Pi_{i=1}^{n} f_{\theta_0}(x_i)},$$

where we agree that for $k \geq n$, $\Pi_{i=k+1}^{n} = 1$, and $\Pi_{i=1}^{k} = \Pi_{i=1}^{n}$.

Define

$$\pi_n = P(\nu \leq n | x - 1, ..., x_n) = \pi_1 + ... + \pi_n.$$

Then, we can see that

$$\frac{\pi_n}{1 - \pi_n} = \sum_{j=1}^{n} l_{j,n}(1-p)^{j-n-1}p,$$

where

$$l_{j,n} = \Pi_{i=j+1}^{n} \frac{f_{\theta_1}(x_i)}{f_{\theta_0}(x_i)}.$$

The Shiryayev–Roberts procedure is defined as the limit of $\pi_n/(p(1-\pi_n))$ as $p \to 0$, which can be written as the following recursive form:

$$R_n = \sum_{j=1}^{n} l_{j,n} = (1 + R_{n-1})e^{(\theta_1 - \theta_0)x_n},$$

for $R_0 = 0$. Thus, the procedure is defined by assuming that the change occurs far away from the beginning.

A change is signaled at the time

$$\tau = \inf\{n > 0 : R_n > T\},$$

where T is the pre-designed threshold.

Given $\tau \geq \nu$, by using the Bayes rule, the posterior distribution for ν can be written as

$$P(\nu = k | \nu \leq \tau; x_1, ..., x_n) = \frac{P(\nu = k | x_1, ..., x_n)}{P(\nu \leq \tau | x_1, ..., x_n)}$$

$$= \frac{l_{k,\tau}(1-p)^{k-1}}{\sum_{j=1}^{\tau} l_{j,\tau}(1-p)^{j-1}}.$$

The posterior mean for ν given $\tau \geq \nu$ is thus

$$\frac{\sum_{j=1}^{\tau} j l_{j,\tau}(1-p)^{j-1}}{\sum_{j=1}^{\tau} l_{j,\tau}(1-p)^{j-1}}.$$

As $p \to 0$, the posterior mean approaches

$$\nu^* = \frac{\sum_{j=1}^{\tau} l_{j,\tau}}{\sum_{j=1}^{\tau} lj, \tau}$$

$$= \frac{Q_\tau}{R_\tau},$$

where

$$Q_n = (n + Q_{n-1})e^{(\theta - \theta_0)x_n},$$

for $Q_0 = 0$.

Thus, we see the estimator ν^* can be calculated recursively by updating two Markovian chains R_n and Q_n simultaneously.

Careful readers may find that there is a difference between the ranges for change-point between the CUSUM procedure and Shiryayev–Roberts procedure. In the CUSUM procedure, we assume that the change-point $\nu = 0, 1, ..., \tau - 1$ and $\hat{\nu}$ only takes the integer values, while here it is assumed that $\nu = 1, ..., \tau$ and ν^* can take any real number between $(0, \tau)$. To make up the difference, we shall take the lower floor $[\nu^*]$, the largest integer less than ν^* as the Bayesian estimator.

10.2 Comparison with CUSUM Procedure

In this section, we conduct a simulation study for the bias of the change-point estimator in comparison with the CUSUM procedure. We select the same parameters $\theta_0 = -0.25, 0.5$. To match with the CUSUM procedure with $d = 10$, we use the approximation for the ARL_0 in Chapter 1 and find the corresponding values for ARL_0 as 2074.28, 141, 347.3 for $\theta_0 = -0.25, -0.5$, respectively. Then we use the approximate formula for ARL_0 for the Shiryayev–Roberts procedure [Pollak (1987) and Wu (1994)]:

$$ARL_0 \approx Te^{-2\theta_0\rho},$$

where $\rho \approx 0.583$. The corresponding values of T are 1550, 78, 903 for $\theta_0 = -0.25$ and -0.5, respectively.

The following table gives the simulated results based on 1000 replications.

Comparing the table with Table 2.1 in Chapter 2 for the CUSUM procedure, we can see that the two procedures make very little difference in terms of the mean of biases. It seems that the CUSUM procedure performs slightly better for $\theta_0 = -0.25$ while the vice versa is true for $\theta_0 = -0.5$.

Theoretical studies for the bias of the change-point estimator and the inference problem for the post-change parameters will be left as future topics.

10.3 Case Study: Nile River Data

Here we use the Nile river data set as an illustration. We shall only concentrate on the change-point estimator.

We standardize the data as for the CUSUM procedure in Chapter 2. The Shiryayev–Roberts process

$$R_n = (1 + R_{n-1})e^{2x_n}$$

is reported as

Table 10.1: Bias of Change-Point Estimator for S−R Procedure

ν	θ_0	θ	$P(N > \nu)$	$E[\nu^* - \nu \mid N > \nu]$	$E[\|\nu^* - \nu\| \mid N > \nu]$
50	−0.25	0.25	0.989	1.647	9.370
		0.5	0.991	−5.387	7.105
		0.75	0.991	−7.648	7.981
		1.00	0.995	−8.360	8.448
	−0.50	0.5	1.000	0.433	2.903
		0.75	1.000	−0.932	2.204
		1.00	1.000	−1.437	2.025
100	−0.25	0.25	0.962	0.498	10.592
		0.50	0.966	−6.941	8.587
		0.75	0.969	−9.634	8.947
		1.00	0.969	−9.090	9.177
	−0.50	0.50	0.999	0.568	2.826
		0.75	1.000	−1.045	2.251
		1.00	0.999	−1.406	1.959

[1] 9.827359e-002 5.691135e-002 1.280628e+000 5.310156e-002
[5] 5.457058e-002 5.464671e-002 1.408634e+001 2.550718e-001
[9] 2.259058e-003 7.152248e-002 7.780850e-001 3.372104e+000
[13] 5.042135e-001 1.109900e+000 1.026999e+000 2.576820e+000
[17] 1.345895e-001 1.895885e+001 2.619773e+001 1.940864e+000
[21] 3.980027e-001 3.255073e-002 6.278947e-002 1.304823e-002
[25] 1.059857e-002 2.005138e-002 4.230999e-001 1.925956e-001
[29] 2.972927e+001 2.664577e+002 1.346093e+003 1.207774e+005
[33] 2.114434e+005 2.050752e+006 1.643843e+008 4.225075e+008
[37] 3.911288e+010 1.903828e+010 5.734221e+009 6.311995e+009
[41] 6.320933e+010 3.396334e+012 1.372119e+016 1.536906e+017
[45] 1.212399e+019 1.191468e+018 1.612476e+017 1.589129e+018
[49] 4.648762e+019 5.463097e+020 1.499071e+022 1.199927e+023
[53] 7.086989e+023 4.321815e+024 3.634623e+026 2.909322e+027
[57] 1.172045e+029 2.054772e+030 7.262688e+029 2.301543e+031
[61] 5.129431e+032 2.981450e+033 2.386492e+034 3.918953e+034
[65] 3.393373e+034 1.182037e+035 1.367049e+036 7.808707e+035
[69] 2.042285e+037 2.442200e+039 4.498432e+041 3.543602e+042
[73] 4.809334e+043 2.000483e+045 3.237506e+046 1.144312e+046
[77] 7.205203e+046 3.626320e+047 2.766639e+048 1.077936e+049
[81] 4.342556e+050 1.614932e+052 1.445865e+053 4.354862e+052
[85] 1.084054e+053 9.091072e+052 1.568505e+054 3.604282e+054
[89] 3.604282e+054 4.662434e+055 2.269450e+055 6.845131e+055
[93] 2.236590e+056 9.876149e+054 2.706179e+055 1.055872e+057

[97] 2.586660e+057 1.579640e+059 1.028423e+061 4.416917e+062

The ratios Q_n/R_n is reported as follows:

[1] 1.000000 1.910520 2.941335 3.405534 4.919601 5.944093
[7] 6.945288 7.015200 8.596623 9.996837 10.933040 11.533101
[13] 11.868615 13.285558 14.098128 15.036398 15.585378 17.713568
[19] 17.778022 17.859719 18.927528 21.125286 22.940900 23.937429
[25] 24.986314 25.989369 26.980134 27.696785 28.789540 28.828931
[31] 28.837049 28.839397 28.839431 28.839456 28.839459 28.839459
[37] 28.839459 28.839459 28.839459 28.839459 28.839459 28.839459
[43] 28.839459 28.839459 28.839459 28.839459 28.839459 28.839459
[49] 28.839459 28.839459 28.839459 28.839459 28.839459 28.839459
[55] 28.839459 28.839459 28.839459 28.839459 28.839459 28.839459
[61] 28.839459 28.839459 28.839459 28.839459 28.839459 28.839459
[67] 28.839459 28.839459 28.839459 28.839459 28.839459 28.839459
[73] 28.839459 28.839459 28.839459 28.839459 28.839459 28.839459
[79] 28.839459 28.839459 28.839459 28.839459 28.839459 28.839459
[85] 28.839459 28.839459 28.839459 28.839459 28.839459 28.839459
[91] 28.839459 28.839459 28.839459 28.839459 28.839459 28.839459
[97] 28.839459 28.839459 28.839459 28.839459

From the numerical results, we see that no matter what value T takes (above 100), the estimator ν^* is always 28 (the largest integer), which is the same as for the CUSUM procedure.

10.4 Concluding Remarks

In this work, we mainly studied the one-variable CUSUM procedure for detecting a change in a one-dimensional interest parameter, mainly the mean. Our main purpose is to provide some techniques and methods to deal with the inference problem for the change-point and the post-change parameter and study the quantitative properties. The methods and techniques are quite general and can be used to study a more general model, such as the time series models and dynamic systems, although the technical difficulty will raise considerably.

A vast amount of literature is available to extend the CUSUM procedure to the multivariate observation case and multidimensional parameter case as well as more general dynamic models. We only mentioned a few references in our discussion and could not be able to track all the literatures available, even in the one-variable case.

The change-point problem is a natural sequential monitoring problem since the change-point is a time index and the main goal is to detect it and to estimate it and post-change parameters after a change is detected. Although a quick detection is vital, the estimation after detection is very important in order to draw the inferences and make a decision.

A few topics should be studied in future research and we list them here.

First, the inference problem for post-change parameters under general exponential family should be studied by directly following the lines in Chapter 4. Typical examples are the change in variance in the stock price case and change in hazard rate in coal mining disaster case. The biases can be studied by taking Taylor expansions, but the confidence sets should be constructed beyond the normal case. A further development should be made when the nuisance parameters are available and subject to change as well, as discussed in Chapter 6.

Second, the time series model can be studied by following the lines in Chapter 9 by using the innovations or residuals. The time-lag and correlation effect as in the AR(1) model case should be studied. The method should be used to study the dynamic systems as well. The readers are referred to Basseville and Nikiforov (1993) and Lai (1998, 2002).

Third, more general models should be studied beyond the exponential family. An example is the change of parameters in ARCH model for the stock price [Schipper and Schmid (2001)]. Recent discussions on models with jumps on both the return rate and the volatility are natural applications. A typical model is given in Duffie, Pan, and Singleton (2000).

Fourth, the multivariate observation case and multidimensional parameter case should be studied as well. Due to the author's limitation to the literature review, we shall not comment too much on this approach, and we leave it to other experts.

Bibliography

Assaf, D. (1997). Estimating the state of a noisy continuous time Markov chain when dynamic sampling is feasible. *Ann. Appl. Prob.*, *7*, 822 – 836.

Avery, P.J. and Henderson, D.A. (1999). Detecting a changed segment in DNA sequence. *Appl. Statist.*, *48*, 489 – 503.

Bagshaw, M. and Johnson, R.A. (1974). The effect of serial correlation on the performance of CUSUM tests. *Technometrics*, *16*, 103 – 112.

Bassiville, M. and Nikiforov,I. (1993). *Detection of Abrupt Changes*. Prentice Hall, Englewood Cliffs, NJ.

Beattie, D.W. (1962). A continuous acceptance sampling procedure based on cumulative sum chart for the number of defects. *Appl. Statist.*, *11*, 137 – 147.

Bertion, J. and Doney, R.A. (1994). On conditioning a random walk to stay nonnegative. *Ann. Probab.*, *22*, 2152 – 2167.

Boys, R.J., Henderson, D.A., and Wilkinson, D.J. (2000). Detecting homogeneous segments in DNA sequence by using hidden Markov models. *Appl. Statist.*, *49*, 269 – 285.

Box, G.E.P., Jenkins, G.M., and Reinsel, G.C. (1994). *Time Series Analysis: Forecasting and Control*. Prentice Hall, Englewood Cliffs, NJ.

Braun, J.V., Braun, R.K., and Muller, H-G. (2000). Multiple change-point fitting via quasi-likelihood, with applications to DNA sequence segmentation. *Biometrika*, *87*, 301 – 314.

Brook, D. and Evans, D.A. (1972). An approach to the probability distribution of CUSUM run length. *Biometrika*, *59*, 539 – 549.

Chang, T. (1992). On moments of the first ladder height of random walks with small drift. *Ann. Appl. Probab.*, *2*, 714 – 738.

Churchill, G.A. (1989). Stochastic models for heterogeneous DNA sequences. *Bull. Math. Boil.*, *51*, 79 – 94.

Coad, D.S. and Woodroofe, M. (1998). Approximate bias calculations for sequentially designed experiments. *Sequential Analysis*, *17*, 1 − 31.

Cobb, G.W. (1978). The problem of the Nile: Conditional solution to a change-point problem. *Biometrika*, *65*, 243 − 251.

Cox, D.R. and Lewis, P.A.W. (1966). *The Statistical Analysis of Series of Events.* Methuen and Co. Ltd. London.

Ding, K. (2003). A lower confidence bound for the change point after a sequential CUSUM test. *J. Statist. Plann. Inf.* , *115*, 311 − 326.

Duffie, D., Pan, J., and Singleton, K.J. (2000). Transform analysis and asset pricing for affine jump-diffusions, *Econometrica*, *68*, 1343 − 1376.

Elliott, R.J., Aggoun, L., and Moore, J.B. (1995). *Hidden Markov Models: Estimation and Control.* New York, Springer.

Fu, Y.X. and Curnow, R.N. (1990). Locating a changes segment in a sequence of Bernoulli variables. *Biometrika*, *77*, 295 − 304.

Fuh, C.D. (2003). SPRT and CUSUM in hidden Markov models. *Ann. Statist.* , *31*, 942 − 977.

Halpern, A.L. (2000). Multiple change-point testing for an alternating segments model of a binary sequence. *Biometrics*, *56*, 903 − 908.

Hawkins, D.M. and Olwell, D.H. (1998). *Cumulative Sum Charts and Charting for Quality Improvement.* Springer-Verlag, New York.

Hines, W.G.S. (1976). A simple monitor of a system with sudden parameter changes. *IEEE Inf. Theo.*, *IT-22*, 210 − 216.

Hinkley, D.V. (1971). Inference about the change point from the cumulative sum test. *Biometrika*, *58*, 509 − 523.

Karl, T.R., Knight, R.W., and Baker, B. (2000). The record breaking global temperatures of 1997 and 1998: Evidence for an increase in the rate of global warming? *Geophysical Research Letters*, *27*, 719 − 722.

Khasminskii, R.Z., Lazareva, B.V., and Stapleton (1994). Some procedures for state estimation of a hidden Markov chain with two states. In *Statistical Decision Theory and Related Topics V* (S.S. Gupta and J.O. Berger, eds.), 47 − 487.

Khasminskii, R.Z. and Lazareva, B.V. (1992). On some filtration procedure for jump Markov process observed in white Gaussian noise. *Ann. Statist.*, *20*, 2153 − 2160.

Khasminskii, R.Z. and Zeitouni, O. (1996). Asymptotic filtering for finite state Markov chains. *Stoch. Proc. Appl.*, *63*, 1 − 10.

Krieger, A.M., Pollak, M., and Yakir, B. (2003). Surveillance of a simple linear regression. *J. Amer. Statist. Assoc.*, *98*, 456 − 468.

Lai, T.L. (1995). Sequential change point detection in quality control and dynamic systems.(with discussion) *J. Roy. Statist. Assoc.* (B), *57*, 613−658.

Lai, T.L. (1998). Information bounds and quick detection of parameter changes in stochastic systems. *IEEE Trans. Inform. Theory*, *44*, 2917 − 2929.

Lai, T.L. (2002). Detection and estimation in stochastic systems with time-varying parameters. *Lecture Notes in Control and Inform. Sci.*,, *280*, *Stochastic theory and control*(Lawrence, KS, 2001), 251 − 265.

Lai, T.L. and Shan, J.Z. (1999). Efficient recursive algorithms for detection of abrupt changes in signals and control systems. *IEEE Trans. Automat. Control*, *44*, 952 − 966.

Lorden, G. (1971). Procedures for reacting to a change in distribution. *Ann. Math. Statist.*, *42*, 1897 − 1908.

Maguire, B.A. , Pearson, E.S., and Wynn, A.H.A. (1952). The time intervals between industrial accidents. *Biometrika*, *39*, 168 − 180.

Moustakides, G.V. (1986). Optimal stopping times for detecting a change in distributions. *Ann. Statist.*, *14*, 1379 − 1387.

Page, E.S. (1954). Continuous inspection schemes. *Biometrika*, *41*, 100 − 114.

Pollak, M. (1985). Optimal detection of a change in distribution. *Ann. Statist.*, *13*, 206 − 227.

Pollak, M. (1987). Average run lengths of an optimal method of detecting a change in distribution. *Ann. Statist.*, *15*, 749 − 779.

Pollak, M. and Siegmund, D. (1985). A diffusion process and its applications to detecting a change in the drift of Brownian motion. *Biometrika*, *72*, 267 − 280.

Pollak, M. and Siegmund, D. (1986). Convergence of quasi-stationary to stationary distribution for stochastically monotone Markov processes. *J. Appl. Probab.*, *23*, 215 − 220.

Pollak, M. and Siegmund, D. (1986). Approximations to the ARL of CUSUM tests. Technical Report, Department of Statistics, Stanford University.

Prabhu, N.U. (1965). *Queues and Inventories*, Wiley, New York.

Reynolds, M.R. Jr. (1975). Approximations to the average run length in cumulative sum control charts. *Technometrics*, *17*, 65 − 71.

Ritov, Y. (1990). Decision theoretic optimality of the CUSUM procedure. *Ann. Statist.*, *18*, 1464 − 1469.

Robbins, H. and Siegmund, D. (1973). A class of stopping rules for testing parametric hypotheses. *Proceedings of 6th Berkeley Symposium of Mathematical Statistics and Probability, 4,* 37 − 41.

Robbins, H. and Siegmund, D. (1974). The expected sample size of some tests of power one. *Annals of Statistics, 2,* 415 − 436.

Roberts, S.W. (1959). Control chart tests based on geometric moving average. *Technometrics, 1,* 239 − 250.

Roberts, S.W. (1966). A comparison of some control chart procedures. *Technometrics, 8,* 411 − 430.

Schipper, S. and Schmid, W. (2001). Sequential methods for detecting changes in the variance of economic time series.*Sequent. Anal., 20,* 235 − 262.

Schmid, W. (1997). CUSUM control schemes for Gaussian processes. *Statist. Papers, 38,* 191 − 217.

Shewhart, W.A. (1931). *Economic Control of Quality of Manufactured Products.* New York, Van Nostrand.

Shiryayev, A.N. (1963). On optimum methods in quickest detection problems. *Theory Probab. Appl., 13,* 22 − 46.

Siegmund, D. (1978). Estimation following sequential tests. *Biometrika, 65,* 341 − 349.

Siegmund, D. (1979). Corrected diffusion approximations in certain random walk problems. *Adv. Appl. Probab.,11,* 701 − 719.

Siegmund, D. (1985). *Sequential Analysis: Tests and Confidence Intervals.* Springer, New York.

Siegmund, D. (1988). Confidence sets in change-point problems. Internat. Statist. Rev., *56,* 31 − 48.

Srivastava, M.S. (1997). CUSUM procedure for monitoring variability. *Comm. Statist. Theo. Meth., 26,* 2905 − 2926.

Srivastava, M.S. and Wu, Y. (1993). Comparison of EWMA, CUSUM and Shiryaye − Roberts procedures for detecting a shift in the mean. *Ann. Statist., 21,* 645 − 670.

Srivastava, M.S. and Wu, Y. (1999). Quasi-stationary biases of change point and change magnitude estimation after sequential CUSUM test. *Sequential Analysis, 18,* 203 − 216.

Stone, C. (1965). On moment generating functions and renewal theory. *Ann. Math. Statist., 36,* 1298 − 1301.

Van Dobben de Bruyn, C.S. (1968). *Cumulative Sum Tests: Theory and Practice.* London, Griffin.

Viterbi, A.J. (1967). Error bounds for convolutional codes and an asymptotically optimum decoding algorithm. *IEEE Trans. Inf.,IT-13,* 260 – 269.

Wald, A. (1947). *Sequential Analysis.* New York, Wiley.

Wasserman, G.S. and Wadworth, H.M. (1989). A modified Beattie procedure for process monitoring. *Technometrics,31,* 415 – 421.

Whitehead, J. (1986). On the bias of maximum likelihood estimation following a sequential test. *Biometrika, 73,* 573 – 581.

Whitehead, J., Todd, S., and Hall, W.J. (2000). Confidence intervals for secondary parameters following a sequential test. *J. Roy. Statist. Soc.*(B), *62,* 731 – 745.

Woodall, W.H. (1983). The distribution of run-length of one-sided CUSUM procedure. *Technometrics, 25,* 295 – 301.

Woodroofe, M. (1986). Very weak expansions for sequential confidence levels. *Annals of Statistics, 14,* 1049 – 1067.

Woodroofe, M. (1990). On stopping times and stochastic monotonicity. *Sequential Analysis, 9,* 335 – 342.

Woodroofe, M. (1990). On the nonlinear renewal theorem. *Annals of Probability, 18,* 1790 – 1805.

Woodroofe, M. (1992). Estimation after sequential testing: A simple approach for a truncated sequential probability ratio test. *Biometrika, 79,* 347 – 352.

Woodroofe,M. and Coad, D.S. (1997). Corrected confidence sets for sequentially designed experiments, *Statistica Sinica, 7,* 53 – 74.

Wu, W., Woodroofe, M., and Mentz, G. (2001). Isotonic regression: Another look at the change point problem.*Biometrika, 88,* 793 – 804.

Wu, Y. (1994). Design of control charts for detecting the change-point. *Change-point Problems,* Lecture Notes 23, Institute of Mathematical Statistics, pp. 330 – 345.

Wu, Y. (1999). Second order expansions for the moments of minimum point of an unbalanced two-sided normal random walk. *Ann. Inst. Statist. Math., 51,* 187 – 200.

Wu, Y. (2004). Inference for the post-change mean detected by a CUSUM procedure. Revised for *J. Statist. Plann. Inf.*

Wu, Y. (2004a). Bias of estimator for the change point detected by a CUSUM procedure. *Ann. Inst. Statist. Math. , 56,* 127 – 142.

Wu, Y. and Xu, Y. (1999). Local sequential testing procedures in a normal mixture model. *Comm. Statist. Theo. Meth.*, *28*, 1777 − 1792.

Yakir, B., Krieger., A.M., and Pollak, M. (1999). Detecting a change in regression: First order optimality. *Annals of Statistics*, *27*, 1896 − 1913.

Yao, Y-C. (1985). Estimation of noisy telegraph processing: Nonlinear filtering versus nonlinear smoothing. *IEEE Trans. Inform.*, *IT-31*, 444 − 446.

Index

absolute bias, vi, 23, 82, 110, 135
adaptive CUSUM procedure, vii, 117, 123, 128, 131
adaptive sequential test, vii, 117
AR(1) model, vii, 130, 134, 141
average delay detection time, 126
average run length, vi, 10

Bayes classifier, 112
Bayesian estimator, 147
bias, vi, 23, 25, 46, 51, 59, 82, 83, 85, 101, 125, 126, 135
boundary crossing time, 3, 122
Brownian motion, 121

change slope, 132
change-point estimator, vi, 2, 31, 48, 64, 101, 117, 125, 134, 141, 147
change-point problem, v, vi, 81
chi-square random variable, 9, 31, 32
classification, vii, 105, 108, 113
conditional expectation, 3, 47, 49, 57, 58, 86
conditional random walk, vi, 67, 70
confidence interval, vii, 37, 57, 82, 89, 95
confidence set, 44
corrected confidence interval, 51, 58, 91
corrected normal pivot, vi, 49, 51, 54, 78, 82, 95
correlation, 139
coverage probability, 38, 40, 43, 58
CUSUM procedure, v–vii, 2, 3, 27, 28, 81, 105, 109, 111, 114, 133, 135, 141, 149

CUSUM process, 16, 28, 30, 34, 37, 52, 83–85, 100

dam process, vii, 104, 105, 109
delay response time, 107, 109, 111, 113
distributional parameter, 133, 134
double CUSUM procedure, 113

Edgeworth expansion, 95
error rate, 105, 106, 108, 110, 112, 113
EWMA procedure, v, 145
exponential distribution, 9, 25, 53, 88
exponential family, vi, vii, 1, 4, 38, 45, 82, 103, 109, 111, 150

false signal, vi, vii, 67, 100
first type error, 109, 113
fixed sample size, vi, 2, 43

geometric distribution, 38, 62

hazard rate, 30
hidden Markov model, 103, 108, 112, 114

ladder variable, 4, 6, 17
linear post-change mean, vii, 117
lower confidence limit, vi, 37

model-based CUSUM procedure, vii, 134, 135, 141, 142
moment generating function, 62

non-linear renewal theorem, 119
normal density, 53, 55, 71, 73, 86
normal distribution, vi, 8, 17, 49

operating characteristics, 2, 133

optimal design, 105, 108, 110, 111
overshoot, 3, 5, 7, 9, 46

post-change mean, vi, vii, 45, 99, 101, 118, 128
post-change mean estimator, vi, 45, 64, 67, 95, 101, 117, 125
post-change parameter, v, vi, 2, 118, 141, 142, 147, 149, 150
post-change variance, vii, 101

quality control, v, vi, 37
quasistationary distribution, 16, 52
quasistationary state, 38, 51, 110, 114, 125, 126

random walk, 6, 16, 17, 32, 45, 46, 60, 68, 83, 85, 86, 135
recursive mean estimation, 126
recursive mean estimator, 119, 120
renewal function, 5
renewal process, 125
Robbins and Siegmund estimator, 119, 121

saddle-point approximation, 95
second type error, 106, 107
second-order approximation, vi, 7, 10, 17, 25, 44, 84, 85, 95, 105, 110, 135
segmentation, vii, 105, 108, 112–114
sequential likelihood ratio test, 2
sequential sampling plan, vi, 17, 25, 82, 95
Shewhart chart, v, 145
Shiryayev–Roberts procedure, v, vii, 145–147
strong laws of large numbers, 59
strong renewal theorem, vi, 4, 5, 10, 83, 110, 115
structural parameter, 134

Taylor expansion, 7, 8, 11, 50, 71, 85, 92, 150
time series model, 133, 149
total probability law, 46, 54, 55, 59
two-sided random walk, 25

very weak expansion, 95

Wald's identity, 3, 60
Wald's likelihood ratio identity, vii, 7, 10, 17, 18, 21, 44, 46, 53, 60, 68, 73, 76, 86, 87, 89, 93, 107, 120, 136
Wiener-Hopf factorization, 7

155: Leon Willenborg and Ton de Waal, Elements of Statistical Disclosure Control. xvii, 289 pp., 2000.

156: Gordon Willmot and X. Sheldon Lin, Lundberg Approximations for Compound Distributions with Insurance Applications. xi, 272 pp., 2000.

157: Anne Boomsma, Marijtje A.J. van Duijn, and Tom A.B. Snijders (Editors), Essays on Item Response Theory. xv, 448 pp., 2000.

158: Dominique Ladiray and Benoît Quenneville, Seasonal Adjustment with the X-11 Method. xxii, 220 pp., 2001.

159: Marc Moore (Editor), Spatial Statistics: Methodological Aspects and Some Applications. xvi, 282 pp., 2001.

160: Tomasz Rychlik, Projecting Statistical Functionals. viii, 184 pp., 2001.

161: Maarten Jansen, Noise Reduction by Wavelet Thresholding. xxii, 224 pp., 2001.

162: Constantine Gatsonis, Bradley Carlin, Alicia Carriquiry, Andrew Gelman, Robert E. Kass Isabella Verdinelli, and Mike West (Editors), Case Studies in Bayesian Statistics, Volume V. xiv, 448 pp., 2001.

163: Erkki P. Liski, Nripes K. Mandal, Kirti R. Shah, and Bikas K. Sinha, Topics in Optimal Design. xii, 164 pp., 2002.

164: Peter Goos, The Optimal Design of Blocked and Split-Plot Experiments. xiv, 244 pp., 2002.

165: Karl Mosler, Multivariate Dispersion, Central Regions and Depth: The Lift Zonoid Approach. xii, 280 pp., 2002.

166: Hira L. Koul, Weighted Empirical Processes in Dynamic Nonlinear Models, Second Edition. xiii, 425 pp., 2002.

167: Constantine Gatsonis, Alicia Carriquiry, Andrew Gelman, David Higdon, Robert E. Kass, Donna Pauler, and Isabella Verdinelli (Editors), Case Studies in Bayesian Statistics, Volume VI. xiv, 376 pp., 2002.

168: Susanne Rässler, Statistical Matching: A Frequentist Theory, Practical Applications and Alternative Bayesian Approaches. xviii, 238 pp., 2002.

169: Yu. I. Ingster and Irina A. Suslina, Nonparametric Goodness-of-Fit Testing Under Gaussian Models. xiv, 453 pp., 2003.

170: Tadeusz Caliński and Sanpei Kageyama, Block Designs: A Randomization Approach, Volume II: Design. xii, 351 pp., 2003.

171: D.D. Denison, M.H. Hansen, C.C. Holmes, B. Mallick, B. Yu (Editors), Nonlinear Estimation and Classification. x, 474 pp., 2002.

172: Sneh Gulati, William J. Padgett, Parametric and Nonparametric Inference from Record-Breaking Data. ix, 112 pp., 2002.

173: Jesper Møller (Editor), Spatial Statistics and Computational Methods. xi, 214 pp., 2002.

174: Yasuko Chikuse, Statistics on Special Manifolds. xi, 418 pp., 2002.

175: Jürgen Gross, Linear Regression. xiv, 394 pp., 2003.

176: Zehua Chen, Zhidong Bai, Bimal K. Sinha, Ranked Set Sampling: Theory and Applications. xii, 224 pp., 2003

177: Caitlin Buck and Andrew Millard (Editors), Tools for Constructing Chronologies: Crossing Disciplinary Boundaries, xvi, 263 pp., 2004

178: Gauri Sankar Datta and Rahul Mukerjee , Probability Matching Priors: Higher Order Asymptotics, x, 144 pp., 2004

179: D.Y. Lin and P.J. Heagerty , Proceedings of the Second Seattle Symposium in Biostatistics: Analysis of Correlated Data, vii, 336 pp., 2004

180: Yanhong Wu, Inference for Change-Point and Post-Change Means After a CUSUM Test, xiv, 176 pp., 2004

181: Daniel Straumann, Estimation in Conditionally Heteroscedastic Time Series Models , x, 250 pp., 2004